Manager's Guide to Mobile Learning

Other titles in the Briefcase Books series include:

Communicating Effectively by Lani Arredondo

Performance Management, Second Edition by Robert Bacal

Manager's Guide to Performance Reviews by Robert Bacal

Manager's Guide to Crisis Management by Jonathan Bernstein

Recognizing and Rewarding Employees by R. Brayton Bowen

Sales Techniques by Bill Brooks

Motivating Employees, Second Edition by Anne Bruce

Building a High Morale Workplace by Anne Bruce

Six Sigma for Managers by Greg Brue

Manager's Guide to Marketing, Advertising, and Publicity by Barry Callen

Manager's Guide to Planning by Peter J. Capezio

Leadership Skills for Managers by Marlene Caroselli

Negotiating Skills for Managers by Steven P. Cohen

Effective Coaching, Second Edition by Marshall Cook and Laura Poole

Conflict Resolution by Daniel Dana

Manager's Guide to Strategy by Roger A. Formisano

Project Management by Gary R. Heerkens

Budgeting for Managers by Sid Kemp and Eric Dunbar

Hiring Great People by Kevin C. Klinvex, Matthew S. O'Connell, and Christopher P. Klinvex

Manager's Guide to Social Media by Scott Klososky

Time Management by Marc Mancini

Manager's Guide to Fostering Innovation and Creativity in Teams by Charles Prather

Manager's Guide to Navigating Change by Stephen Rock

Presentation Skills for Managers by Jennifer Rotondo and Mike Rotondo Jr.

Finance for Non-Financial Managers by Gene Siciliano

The Manager's Guide to Business Writing, Second Edition by Suzanne Sparks Fitzgerald

Skills for New Managers by Morey Stettner

Manager's Survival Guide by Morey Stettner

The Manager's Guide to Effective Meetings by Barbara J. Streibel

Managing Multiple Projects by Michael Tobis and Irene P. Tobis

Manager's Guide to Online Marketing by Jason D. Weaver

Accounting for Managers by William H. Webster

To learn more about titles in the Briefcase Books series go to www.briefcasebooks.com

Manager's Guide to Mobile Learning

Brenda J. Enders

New York Chicago San Francisco Athens
London Madrid Mexico City Milan New Delhi
Singapore Sydney Toronto

1 2 3 4 5 6 7 8 9 0 QFR/QFR 1 8 7 6 5 4 3

ISBN 978-0-07-181147-7
MHID 0-07-181147-8

e-ISBN 978-0-07-181148-4
e-MHID 0-07-181148-6

This is a CWL Publishing Enterprises book developed for McGraw-Hill by CWL Publishing Enterprises, Inc., Madison, Wisconsin, www.cwlpub.com.

This book is printed on acid-free paper.

Contents

Acknowledgments

First, I would like to thank John Woods of CWL Publishing Enterprises for asking me to write this book and giving me the opportunity to share not only my experiences and insights but also what I've learned from other industry experts. Mobile learning is definitely not a one-size-fits-all approach. I hope the insights I share with you will aid you in managing your mobile learning projects.

I also would like to thank those in the mobile learning industry who have helped me to frame my views on mobile learning. While there are too many to list, I would like to extend a special thanks to David Metcalf, Clark Quinn, Judy Brown, Gary Woodill, Robert Gadd, and Chad Udell who, through their books, blogs, conference presentations, and conversations, have had a profound influence on my views of mobile learning.

The other individuals without whose support I wouldn't have been able to write this book include my family. First, thanks to my husband, Ken Allen, for his continuous support, understanding, and for ultimately being my biggest fan. I'd also like to thank Patti Quinn and Tom Enders who gladly listened when I needed a sounding board and nontechie views. Your words and support helped more than you know. Last, a special thanks to my father, Lee Enders, who has unconditionally supported me in my decisions throughout life and encouraged me to follow my dreams.

I would like to dedicate this book to my mother, Dorise Enders, who, before passing away in June 1999, left me with this final piece of advice that I believe is worth sharing with everyone: "You can do anything you put your mind to." I hope this advice helps you as much as it has helped me over the years.

Introduction

As I finish writing this book, I find myself reflecting on the past. A lot has changed over the past two decades in both the learning and development field and my own professional career. When I started almost 20 years ago, I was working for a vendor as an end-user software trainer. Basically, I would train daylong classes on the latest technology, which happened to be the personal computer running Windows 95, the early versions of office productivity software (Word, Excel, and PowerPoint), and even the Internet. Yes, you read that correctly—the Internet, and please stop laughing. It's 100 percent true. Back in the day, many companies would provide training to their employees on the Internet—not how to create web pages or write code but actually explaining what the Internet was all about such as how to search for information and download files.

As with most classroom trainers, I really enjoyed seeing that moment when something you shared with a student made a difference. You could see it in their faces. We'd always say, "You can see the lightbulb come on when they got it." When I trained on how to use the Internet, it went beyond having that aha moment. For many, you could tell they were awestruck by the potential and how the Internet had the possibilities to fundamentally change how they worked. For others, you could see they were curious but at the same time resistant, since it was so foreign to them. I remember some even commented that the Internet would never last, that it was just the technology fad of the moment. Boy, weren't they wrong.

Since my days in the classroom, I've been blessed to be involved in the design and development of a wide variety of learning solutions and even a few award-winning projects. I've experienced hundreds of learning solutions, and I've never seen people react the same as they did when they were first exposed to the Internet—that is, until I started showing people mobile learning solutions.

Many can immediately see the benefit of having anytime, anywhere access to learning opportunities and can envision uses in their company. For others, they imagine the same old e-learning courses, only displayed on a smaller screen with maybe a concern over security.

Whether you fall into the group of people who can see the potential or you find yourself on the fence, this book will guide you along the path to effectively instituting and managing mobile learning projects. I see this book as my opportunity to not only open your eyes to the potential of mobile learning but also to provide you with the tools necessary to see your mobile learning initiatives succeed.

In the first four chapters of the book I focus on providing you with an overview of mobile learning. Think of this as the big picture section of the book. Chapter 1 focuses on the state of the mobile learning industry, including adoption rates and how mobile technologies are changing our personal and business lives. Chapter 2 explains the business drivers that support mobile learning, the principal benefits of leveraging mobile learning, and the challenges you may face when managing your project. Chapter 3 shows the true power of mobile learning, including what drives employees to learn on the job and how this impacts mobile learning design. Chapter 4 brings it all together with a variety of mobile learning practices so you can provide your employees with information (podcasts, videos, e-books, mobile apps), gather learner data (assessments, surveys, polls), and facilitate communication and collaboration among the project participants. You'll also read about some cutting-edge and innovative approaches to mobile learning, including gaming, augmented reality, and immersive simulations.

Chapters 5 through 7 focus on defining and documenting your mobile learning strategy and business plan. These chapters provide a framework for creating and documenting your mobile learning strategy, gathering requirements, and documenting your project's technical and environmental considerations. Chapter 8 helps you decide whether to use an internal development team or partner with an outside vendor to bring your vision to life.

With a solid plan in place, the final three chapters focus on managing the project. We cover working with your stakeholders and development team and the tactics you can use to minimize project risks and increase the odds of a successful mobile learning endeavor.

Chapter 9 focuses on gaining stakeholder support and providing approaches to ensure their goals are being met and project risks are minimized. Chapter 10 moves into how your development team will bring your mobile learning solution to life. The text also gives you some insights into the development process and key milestones that will need your attention. Chapter 11 offers practical advice on techniques that will increase the odds of a successful project. We discuss running a pilot program for your mobile learning solution and the insights you can gain through this process. We review ways to ensure seamless project communication and tips for selecting the right project manager for your mobile learning initiative.

Special Features

Titles in the Briefcase Books series are designed to give you practical information written in a friendly, person-to-person style. The chapters deal with tactical issues and include lots of examples. They also feature numerous sidebars that give you different types of specific information. Here's a description of the sidebars you'll find in this book.

Every subject has some jargon, including the mobile learning field. The Key Term sidebars provide definitions of terms and concepts as they are introduced.

The Smart Managing sidebars do just what their name suggests: give you tips to intelligently apply the strategies and tactics described in this book so you can effectively implement your mobile learning initiatives.

Tricks of the Trade sidebars give you insider how-to hints on techniques astute managers use to execute the tactics described in this book.

It's always useful to have examples that show how the principles in the book are applied. The For Example sidebars provide illustrations, along with case studies of the successful use of mobile learning in various contexts.

Caution sidebars warn you where things could go wrong when implementing your mobile learning initiative.

How can you make sure you won't make a mistake when you're trying to implement the techniques the book describes? You can't, but the Mistake Proofing sidebars give you practical advice on how to minimize the risk of this happening.

The Tools sidebars provide specific directions for implementing the techniques described in the book in a systematic fashion.

Manager's Guide to Mobile Learning

Mobile Learning Overview

When I speak to audiences about mobile learning, I often jokingly say that our phones and other mobile devices have become part of our DNA. Think about it for a minute. We never leave home without them. They are always on. We check them an average of 34 times per day; panic if we lose them; and a large percentage of us sleep with them within an arm's reach. Let's face it: mobile devices have become a part of who we are and how we interact with the world around us

Mobile Device Adoption Rates

It seems that everywhere we look, people are on their mobile phones and tablets, but how many people actually have mobile subscriptions? According to the International Telecommunication Union (June 2012), at the end of 2011, there were approximately 6 billion mobile subscriptions. With the world's population standing at about 7 billion, 6 billion subscriptions represents more than 85 percent of the world's population. Wow! So does that mean almost everyone in the world has a mobile phone? Not quite. Many individuals have multiple subscriptions, and Ericsson (June 2012) believes the actual number of unique subscribers as of the first quarter of 2012 was about 4.2 billion. So we are looking at close to 60 percent of the world's population that owns and uses cell phones.

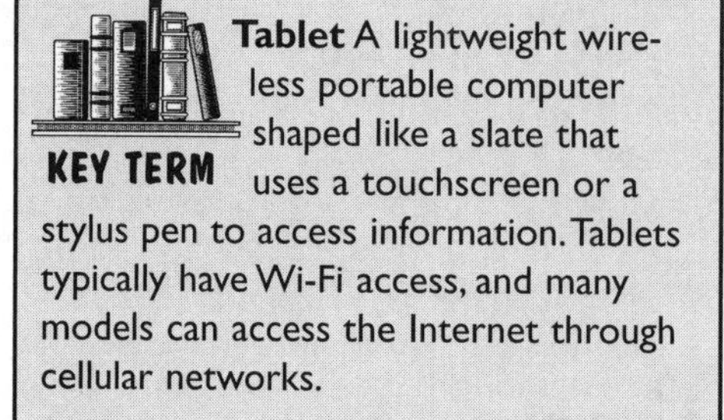

Tablet A lightweight wireless portable computer shaped like a slate that uses a touchscreen or a stylus pen to access information. Tablets typically have Wi-Fi access, and many models can access the Internet through cellular networks.

With the ever-growing global workforce, I'd like to take a moment and look at how mobile devices have been adopted worldwide by region.

Portio Research (see Table 1-1) has broken the mobile penetration landscape into five worldwide regions, and, based on its 2011 data, the only regions that didn't exceed 100 percent penetration were Asia Pacific, Africa, and the Middle East. They forecast by the year 2016 the only regions not to exceed 100 percent would be Africa and the Middle East with 91.3 percent.

Region	Mobile Penetration (in Percent)							
	2009	**2010**	**2011**	**2012F**	**2013F**	**2014F**	**2015F**	**2016F**
Europe	125.1	127.4	131.4	133.6	135.4	136.8	137.8	138.5
Asia Pacific	55.7	67.1	77.6	85.1	92.0	97.9	103.1	107.4
North America	92.1	98.1	102.6	106.7	110.3	113.4	116.0	118.1
Latin America	84.7	93.0	100.4	106.7	110.9	113.5	115.5	116.6
Africa/ Middle East	56.1	65.0	71.2	77.1	81.8	85.7	88.9	91.3
Worldwide	68.5	77.8	86.1	92.3	97.7	102.3	106.1	109.2

F = Forecast

Source: Portio Research Ltd.

Table 1-1. Projected regional mobile penetration (in percentages, 2009–2016)

Not only are we looking at more than half the world with mobile subscriptions in 2012, and a future growth rate to exceed 100% worldwide by 2014, but mobile technologies (including devices and mobile software) are the fastest-growing technology in history to date.

Data Plan Costs

Currently in the United States, mobile carriers are doing away with the unlimited data plans. At the time of this writing, Sprint is the only major U.S. carrier that still offers an unlimited data plan to new subscribers. This leads to an implication that data plan costs are skyrocketing. In reality, the cost per MB of data has continuously dropped since 2008. Portio Research forecasts this trend will continue, and by 2015, the cost will be a mere US$0.01 per MB (see Figure 1-1). Free Wi-Fi networks are popping up everywhere—from your local McDonald's to your local gas station. The problem is not the cost of the service, but rather, most individuals purchase a data plan that exceeds their needs, as the typical U.S. mobile subscriber uses less than 2GB per month.

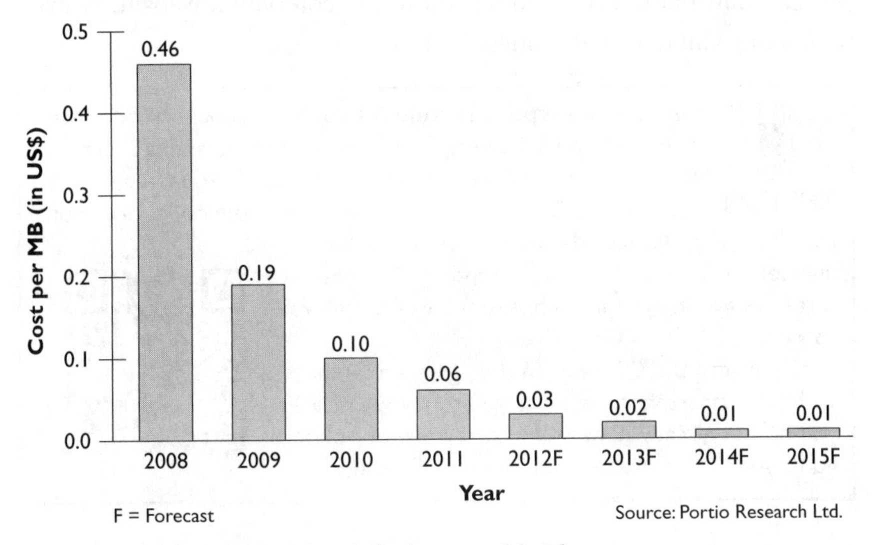

Figure 1-1. Cost per MB of mobile data, worldwide

Corporate Adoption

All this sounds great. Most people are hooked on their mobile devices, and data costs are dropping. But will corporations adopt this technology, or is it all only for personal use? According to Gartner Research, by 2014 it will be commonplace for organizations to support corporate applications on employee-owned smartphones and mobile devices.

Smartphone A cell phone that runs on a mobile computer operating system. While features vary, they typically include a full keyboard (slide out or touchscreen), high-resolution screen, high-speed Internet connectivity, GPS location services, camera, video, and third-party applications (apps).

In addition to supporting enterprise-wide corporate applications, organizations have been using social media and mobile technologies as a way to reach their consumers. For example, if you "like" a company's Facebook page, the company may send you coupons for discounted or even free products. Flip through the pages of any magazine or newspaper, and you will see quick response (QR) codes, which, if you scan them with your smartphone, will give you additional product information by sending you to the company's website or even redirecting you to watch a video.

QR (quick response) code A two-dimensional barcode that can store up to 4,296 alphanumeric characters and is accessed from your smartphone and some cell phones. You need both a camera and QR reader application to scan the code. Common uses for QR codes include sharing company phone numbers or text-based content, sending SMS messages, connecting you to websites, and redirecting you to videos.

If you scan the QR code in the figure shown here with your smartphone, it takes you to my website. If you do not already have a QR reader on your phone, go to your app store and search for "QR reader."

With the information I've shared with you in this section, it's easy to see that mobile technologies aren't only the hot devices of the moment—they're here to stay. Corporations are quickly (or not) adapting their IT strategies to include mobile technologies as a way to support their internal applications as well as their customer needs. As managers, we must explore how we can leverage this technology to aid in training employees, enable employees to achieve greater performance, and educate our customers.

What Is Mobile Learning?

While we cover this question in detail in Chapters 3 and 4, I'd like to briefly take a look at mobile learning, also referred to as *m-learning*, from the 50,000-foot view.

> **KEY TERM**
>
> **Mobile learning** Any activity that allows individuals to be more productive when consuming, interacting with, or creating information. Mediated through a compact digital portable device that the individual carries on a regular basis, has reliable connectivity for, and fits in a pocket or purse. As mobile learning is a developing field, the definitions are numerous. This definition (which is my favorite) comes from The e-learning Guild's "360° Research Report" (2007).

Mobile learning isn't really about the technology or the device, and it shouldn't be a primary mechanism to deliver your e-learning courses. Rather, think of mobile learning as a way to provide your employees or customers with the right content at the right point in time. Consider allowing them and, indeed, encouraging them, to create their own content to support peer-based and social learning opportunities. Look for ways to maximize the devices' unique attributes to augment and extend the learning process. Last, support your learners on the job with performance support tools accessible from their mobile devices. Make your m-learning tools a prominent component of your corporate teaching tool kit and part of a well-thought-out, long-term employee and customer support strategy.

Devices Used for Learning Today

Today, a variety of mobile devices deliver mobile learning experiences, and the lines between these technologies is blurry. While many of us have a variety of mobile devices close at hand, soon we will own only one that accomplishes all our goals. Let's take a moment to look at some of the technologies that deliver mobile learning experiences today.

Portable Digital Assistant

The *portable digital assistant* (*PDA*), a long-time staple for many executives, is on the verge of extinction. They are often called a *palmtop computer*, and their primary focus is on personal organization and office productivity. PDA software typically includes a calendar, address book, notepad, e-mail, and picture and/or video display. PDAs contain memory cards to save data and use Wi-Fi or Bluetooth technology to connect to the Internet. Since they do not connect to cellular networks, their Internet accessibility is tremendously limited compared to other, more modern technologies.

Feature Phones

I often refer to a feature phone as a *dumb phone* since it does not run on a mobile computing platform. Typically feature phones are used to communicate, either via voice or SMS. These phones are often overlooked as a delivery mechanism for a mobile learning experience due to their low-tech feel. However, as we discuss later in the book, SMS is a viable way to distribute learning content to your employees and customers, so do not dismiss feature phones as a potential delivery vehicle.

KEY TERMS

Feature phone A cell phone that includes voice service and the ability to send and receive text messages. It also incorporates features such as a slide-out keyboard, web browsing, and a camera.

SMS This stands for *short message service*, which allows text messages (typically up to 160 characters maximum, although some systems support up to 224 characters) to be sent between phones on any cellular network. If the user's mobile phone is not on when the message is sent, the system holds the message until the phone is turned on.

Portable Media Player

The portable media player isn't just for listening to your favorite music. Many organizations are creating podcast series as a way to deliver learning content to their audience. By leveraging a subscription service, the company can post new content to the Internet for employees (or customers) to listen to or view at their convenience—whether that's on the subway, in the gym, or while you're waiting for your next appointment.

Smartphone

Thanks to the smartphone, no matter where you are, you can check and sync your e-mail from multiple accounts, listen to music, take and share pictures and/or video, utilize GPS location services, and download applications for pleasure and business. While the majority of existing phone owners have a feature phone, that is quickly changing. According to Nielsen, as of February 2012, nearly 50 percent of U.S. mobile owners have smartphones, and more than two-thirds of new phone buyers in 2012 opted for smartphones over feature phones. Smartphones have a tremendous potential to change the way companies deliver training to employees, both as a communication tool and as a user-generated content-creation tool.

> KEY TERM
>
> **Podcast** A multimedia file (audio, images, and/or video) made available on the Internet, typically through a subscription service, for downloading to a portable media player. The subscription feed of automatically delivering new content is what distinguishes a podcast from a simple download or real-time streaming.

Tablets and e-Readers

I think of tablets as the bridge between the smartphone and the laptop. Besides having many of the capabilities of a PC, most tablets include wireless Internet browsing functions, potential cell phone functions, GPS navigation and location services, video camera functions, and a battery life of three to ten hours. The tablet's larger screen size, touchscreen technology, and long battery life make it one of the most interesting devices for creating highly interactive mobile learning. A subset of the tablet is the e-reader, such as the Kindle, used primarily to store and read books. The power of the e-readers is the very large amount of text-based content that can be held on one device, which is accessible from any location with a screen size designed specifically for reading.

> **e-Reader** A device with wireless connectivity to download and read books, newspapers, magazines, documents, and pdf files.

Handheld Gaming System

Handheld gaming systems, such as the Nintendo DS or Sony PSP, can be used to deploy highly interactive serious learning games that allow us to learn from our mistakes in a safe environment. This all-in-one portable device has a built-in screen, traditional game controls, speakers, and replaceable or rechargeable batteries. As with other technologies discussed here and the high adoption rate of smartphones, this technology may be at risk of extinction, at least as a learning platform.

Myths vs. Realities of Mobile Learning

There are many misconceptions or myths about mobile learning. Unfortunately, many managers allow these myths to limit their ability to see and envision the power of mobile learning. So let's address some of the most common myths.

Myth: Screen and Keyboard Are Too Small

At the heart of it, this myth assumes all mobile learning takes place on a mobile phone. As you just read, there are a variety of mobile devices you can leverage to deliver effective mobile learning. Sure, mobile phone screens are significantly smaller than other traditional devices, but if they had large screens, we wouldn't have them with us wherever we go.

Whenever I hear the screen is too small, my first response is always "Too small to do what? Too small to take a traditional e-learning course?" Yes, I agree wholeheartedly. Remember, mobile learning isn't an alternative delivery method for your e-learning. Sure, organizations are converting their existing e-learning libraries to mobile versions, but typically this is to allow users the choice of how they would like to take the course.

Myth: We Can't Provide Devices to All Our Employees

While there are times when purchasing devices is the best answer, such as tablets to be used for sales training or new employee orientation programs, most companies can't justify the expense, and that's okay. Probably most of your employees or customers already have some sort of mobile device that they personally selected, are comfortable using, and best of all, already carry with them everywhere they go. So why worry about providing the devices? Consider allowing your employees to bring their own device (BYOD), and you develop the learning experiences to run on multiple platforms.

Myth: Mobile Learning Must Be Highly Interactive

There's a time and a place for high levels of interactivity, but not all mobile solutions are or should be highly interactive. One benefit of mobile learning is the ability to address learners at their moment of need. For example, videos are not interactive. However, they are beneficial for observing how a procedure is performed prior to actually completing the task. Another example of effective passive learning is having documents, manuals, or job aids on the devices.

> **PASSIVE LEARNING IN ACTION**
>
> FOR EXAMPLE
>
> For years, I've watched people wakeboard at the beach and have longed to learn how. Unfortunately, I had one tiny barrier: I didn't know anyone who could teach me. During a weeklong vacation at Mark Twain Lake, I decided I wasn't getting any younger and thought I'd give mobile learning a shot. I grabbed my iPad, connected to my cellular network, and searched the Internet for "learning to wakeboard." I found many videos and websites on this subject. I began watching YouTube videos to see the basics in action. During my search, I also found a website listing 10 basic steps for wakeboarding and promptly e-mailed a link to myself for future reference.
>
> Sitting in the back of a boat, while we searched for a quiet spot on the lake, I began studying those 10 steps on my my iPhone. Soon we found a spot to give it a shot. With my board strapped on and floating on the water, I nervously ran through the steps in my mind one last time. It's now or never, I thought, so I yelled, "Hollywood" (our code word for *ready*). The boat was moving, and to my amazement, I was up and cutting through the water like a hot knife slicing through a stick of butter. I felt like a kid and kicked myself for not thinking about mobile learning as a way to learn sooner.

Myth: Mobile Learning Must Be SCORM Compliant

SCORM compliance is a complicated, and honestly, a tricky subject when it comes to mobile learning. Most managers don't really care about SCORM and simply want the ability to gather

> 
>
> KEY TERM
>
> **Sharable Content Object Reference Model (SCORM)** A set of standards and specifications for e-learning that define communications between the learning content objects and the learning management system.

solid tracking and measurement data of their learning initiatives. Luckily, most mature mobile learning tools allow for tracking data to be saved without dealing with SCORM, and later it can be integrated with your other learning systems as well as other internal systems.

Myth: Mobile Learning Is Time Consuming to Develop and Costly

When I hear this myth, right away I ask, "What type of mobile learning experience are you creating, and what intent does it serve?" While serious games and outside developed applications can be costly, that isn't the case for most solutions. Many tools are available to develop and deliver cost-effective solutions, and as the field continues to mature, this trend will only continue. This myth also assumes that all content will be generated by the IT or HR department. Consider for a moment the benefit of having key salespeople record videos from their smartphone, sharing their top five tips for success. Keep in mind: mobile learning does not have to be professional production quality and high fidelity sound to be effective. Peer-to-peer learning on a mobile device can be one of the most effective ways to deliver great content.

Myth: People Don't Want Mobile Learning Experiences

This myth can be debunked simply by observing any public setting. The number of people who use the mobile devices as more than a communication tool is staggering. You see people reading books, playing games, reading a newspaper, searching for information on the Internet, getting their e-mail, listening to music, or interacting on a social network service such as Facebook or Twitter. Why would we not want to use that same technology as a learning tool, delivering content when and where we need it? Most people like to attend a classroom or seminar for the social aspects or to get out of work for the day and are not driven to that method for the great learning experience. Let's face it. We are a society that has come to expect mobile experiences, and our learning programs need to incorporate this element.

LEARNING AT THE MOMENT OF NEED

FOR EXAMPLE

My friend Ryan recently came back from his first trip to New York City raving about the public transit system. I had a flashback to my first experience on the NYC subway. I wondered, "Which color train do I need to take? When do I need to change trains? Am I even going in the right direction?" While I love the convenience, low cost, and speed of the subway, it isn't necessarily the easiest system for an outsider to learn. I asked Ryan if he had any issues navigating the subway. He looked puzzled and responded, "Why would I? I have an iPhone, and there's an app for that." I thought to myself, "I learned through the school of hard knocks, and Ryan learned by using his smartphone and an app designed to teach him in his moment of need."

Manager's Checklist for Chapter 1

- ☑ Mobile technologies (devices and supporting software) are the fastest growing technology in history. More than 60 percent of the world's population currently has subscriptions, and data plan costs are dropping.
- ☑ Mobile learning isn't simply loading your e-learning courses on a mobile phone.
- ☑ Mobile learning is about taking advantage of the unique attributes and portability of the devices to create new and unique learning experiences.
- ☑ Understand the potential barriers to mobile learning, but don't let myths and misconceptions stop you from leveraging this powerful tool in your learning strategy.
- ☑ The lines between devices to deliver mobile learning are blurring. In the future, most current devices will be replaced by one device that does it all.

The Time Is Now for Mobile Learning

As a society, we have embraced mobile technologies. Take a moment and think about how mobile technologies have improved your life. Possibly you're staying in touch with family and friends by posting status updates and uploading photos on Facebook. Maybe you're saving money by finding the gas station in your area with the cheapest gas prices. You may be using small, precious moments of free time to make reservations at the newest hot restaurant in town and then having your phone give you turn-by-turn directions to get there.

It's safe to say that as a society, we are hooked on mobile devices and the productive benefits we derive by using them. If you have not already started to think about how to leverage this power in your organizational training strategy, now is definitely the time to start.

Other than personal adoption, why should we act now to incorporate mobile learning in our training programs? In this chapter, we address key data that support the assertion that *now* is the time to act. We answer questions such as: Why should companies implement mobile learning throughout the organization? What are the key business drivers for adding mobile learning to your training strategy? What unique benefits can be achieved through mobile learning initiatives? And last, we address some of the key challenges to consider when implementing mobile learning solutions. Let's start by looking at the business drivers favoring the use of mobile learning in your organization.

Business Drivers for Mobile Learning

When I think about what motivates us from a business perspective to implement mobile learning, three points immediately come to mind. First, mobile learning can be a way to better support our modern workforce. Second, mobile learning can improve your employees' job performance. And third, mobile learning can increase the impact of your training throughout the organization. Let's take a look at each of these business drivers. This all results in what we might call the mobile workforce.

Supporting the Modern Workforce

As times change, so do our employees. We must constantly adapt the work environment to remain successful. One of the largest changes and challenges in today's work environment is the entry of the Millennial generation and how their work style differs drastically from the previous three generations that are currently in our workforce. In the United States, we often refer to the four demographic groups as the Traditionalists (or Silent Generation), Baby Boomers, Gen X, and—the most recent addition to the workforce—the Millennial generation, also referred to as Gen Y. Here is a list of the generations and years of birth that identify them.

Generation	Born
Traditionalist	1928–1945
Baby Boomer	1946–1964
Gen X	1965–1980
Millennial	After 1980

With Millennials comprising approximately 25 percent of today's workforce and the percentage growing each year, we're faced with a substantial group of people who have grown up in a world very different from the other generations and who are technology savvy. Millennials have not only lived their entire lives immersed in technologies on a 24/7 basis, they embrace the constant evolution of those technologies.

Think about it: this is the generation that most likely has never listened to or seen a 45 rpm record and has never spent hours creating a mixed tape of special songs for their boyfriend or girlfriend. Their parents didn't give them the talk about always having coins in case an emergency arose and they needed to use a pay phone to call for help. Instead, this generation has always listened to perfectly mastered DVDs without the static of a record needle. They quickly search for music on the Internet and download it to create a digital playlist they can share with their friends, not having to pause the tape while they shuffle through records and cassettes to find the beginning of the next song they want to record. There is no need for them to have change for a pay phone. There are hardly any pay phones left, and they all have their own cell phones that

Embrace the Generational Differences

I often say the Millennials are wired differently from the rest of us. And research is now showing us the truth of this statement. Their brains are evolving due to the constant influences of technologies. It's common in the workplace to hear managers make statements such as: "Employees need to do it our way" or "That's not the way things are done here." I challenge managers to think differently about this when dealing with Millennials. To do so managers need to become familiar with this generation's life experiences and how to effectively prepare them to excel in their work lives. Their lives have been saturated in technology from Day One, ranging from the way they play to the way they learn to the way they interact with others. This is the generation that does not write; they keyboard. This is the generation that is in constant multitasking mode, flipping the switch to go from one task to another at a moment's notice. This is the generation whose friends are no longer limited by the proximity of their neighborhoods, schools, or even cities. Their social networks are their lifelines, and Millennials do not hesitate to reach out and ask their network how to solve problems, including work problems.

This is the generation that expects their work environment to be at least as tech savvy as they are. If we don't create a work environment that embraces and provides tools to support this generation's professional growth, they will move to another organization that does. When designing your learning solutions, we urge you to not only understand how the Millennials are "wired" but to include them in the solution's design and development process. Ask them how they would tackle the problem; ask them what tools would benefit them and why; include them as part of your testing group. I'm not suggesting that we design our learning solutions specifically and only for them, but rather to take cues from what's going on and create solutions that your whole workforce can learn from to improve your company's performance and productivity.

never leave their pockets. They rarely use landline phones to make calls. Texting is their primary means of communicating.

Mobile Workforce

The modern workforce is not only about the Millennials; it's also about how all of us have changed the way we work and when and where. The days of going to the office from 9 to 5 with a break for lunch are long gone for most employees.

Being an independent consultant, I am exposed to a variety of work environments. I have observed three common ways employees are working today. Many no longer go to the office daily. Instead they work from remote locations such as their home and go into the office only when face-to-face is the best way to accomplish their goals. A second group of employees blends time between the corporate office and time on the road, at a client site, and possibly working from a remote office. A third group of employees are those who work most of their hours in a traditional corporate setting; however, they spend time away from the office, staying connected via cell phone and e-mail.

In today's competitive environment, we are charged to do more with less. The mobile workforce comes in many flavors, but one thing is for sure: it's here to stay. According to International Data Corporation's (IDC) "Worldwide Mobile Worker Population 2011–2015 Forecast," approximately 75 percent of the North American workforce used smartphones, tablets, or both on a daily basis in their jobs in 2010. While the United States is leading the pack with a mobile workforce, IDC forecasts that the world will reach 1.3 billion mobile employees by 2015. This number represents one-third of total enterprise employees worldwide. Those are staggering numbers, and as managers, it's our job to find ways to support our employees regardless of when or where they are working.

Improving Job Performance

We hear it all the time today. To be competitive in today's marketplace, we need to find ways to increase our productivity. But in today's hectic world, one common thread is the need for more time. We are juggling work lives and feeling the pressure to do more in less time. In our busy home lives, we are often chauffeuring kids from one activity to the next, volunteering, working out, and maintaining our homes. The line between work lives and home lives is blurry, and we often grasp small moments to check items off our to-do lists.

Let me share a personal story.

I was heading out of town for a learning conference in Las Vegas where I was to speak to a group of human resource professionals about "Addressing the Needs of the Millennials in the Workforce." It was 3:30 in the morning and I had barely finished my first cup of coffee. My ride arrives to take me to the airport and, unfortunately, we run into more traffic than we anticipated. While sitting in the backseat, my mind begins running through my last-minute to-dos. My first thought is checking into my flight. My meetings ran late the day before, and my evening was filled with family activities. My original plan was to check in when I got to the airport, but now time is of the essence. I pick up my iPhone and launch the American Airlines application. I not only check in to my flight but choose to have my boarding pass stored on my phone. While I'm in this app, I check the First Class upgrade list to see how many people are ahead of me. I'm second, so there's a good chance I can snag a free upgrade. Great. Now I won't need to stop after going through security to pick up a breakfast snack and a bottle of water. Okay. I'm done with everything airport related.

Now I need to check the shortest route from the airport to the hotel. The cabbies in Las Vegas have a bad reputation for taking you the most expensive route if you don't specify otherwise. Phone in hand, I google the route and find that I need to tell the cabbie to go to the hotel via Paradise Road. Then I receive a text message from American Airlines telling me that my gate has been changed, but the flight is still on time. A few minutes later I arrive at the airport and head to security. I scan my electronic boarding pass, show my ID, and board the plane. While the other passengers are boarding, I use a mobile app to secure a reservation for a table of six at 7:45 p.m. As the flight attendant announces that the door has been closed and we are to shut off all electronics, I click the Send button on an e-mail to let my dinner companions know the details of our evening plans.

While in flight, I thought about how much I had accomplished before takeoff: boarding pass, gate change, route to my hotel, dinner reservations, e-mail to dinner companions. This was all time that until recently would have been unproductive.

Improving performance in the context of mobile learning is not only about turning those downtime moments into productive time. It's about finding ways to provide our employees with tools and support to perform tasks more quickly at a high-quality level, and provide them at the specific moment they most need them. How can you leverage mobile technologies to accomplish this goal? We discuss this in detail in Chapter 4, but for now I think we can agree that we must strive to improve our performance to be competitive, and mobile technology is one great way to do that.

Increasing the Impact of Your Training

We all need our training programs to have the maximum positive impact on our organization. We are constantly challenged to create not only an engaging but also an effective training environment so learners can quickly apply what they have learned within the context of their job.

One way we can increase the impact of training is by using the mobile devices as a way to supplement our weaknesses. What exactly do I mean by that? Let's look at one of the most common weaknesses in people: our memory. Specifically I'm referring to our ability to recall facts and lists of information. There are some people who have the knack of remembering this type of information, but for the majority of us, it is definitely not a strong suit. Mobile devices, however, are designed to store and quickly recall information at the touch of a button. Think about it. How many times do we provide lists of information that we need our learners to apply on the job? What type of impact on training could you envision if your organization provided some mobile performance support when you need your employees to not only remember but also apply those lists of facts or tasks in the context of their jobs?

Other key ways mobile learning can impact the overall effectiveness of your training is by:

- Increasing learner engagement and promoting active learning;
- Increasing retention and application of content by spacing the content distribution over a period of time versus having the learner absorb it all at one time;

- Building a sense of community among your employees to share best practices;
- Encouraging learners to reflect on their learning opportunities through social learning and/or user-generated content; and
- Providing learners with context and the ability to quickly apply the knowledge they have gained.

Benefits of Mobile Learning

Now that we've discussed the three main business drivers to implementing mobile learning (supporting the modern workforce, improving performance, and increasing the impact of your training program), let's look at the benefits of mobile learning. When thinking about the benefits of mobile learning, we focus on the unique attributes and characteristics the mobile device brings to the equation. Before reading more, take a moment and think about what you can accomplish on a mobile device that you can't do on your computer or in a classroom setting.

Anytime/Anywhere

Because our mobile devices are not only portable but they are almost always with us, we have the opportunity to provide content any time of the day or night and anywhere the learner wants it. For example, this could be at 7:00 in the evening while you are at a restaurant enjoying a quick dinner or at 5:30 during your morning gym workout, or it could even be 11:15 in the morning while you are preparing for a sales call that starts in 15 minutes. Anytime/anywhere is a great way to give your employees control over when they participate in their training experience. The anytime/anywhere quality gives employees the opportunity to take advantage of those downtimes that we all experience and turn them into productive moments in time.

Just in Time

Taking the anytime/anywhere benefit a tad further, we find the just-in-time benefit. Whenever I hear this phrase, I can't help but imagine a superhero flying in to save the day and, honestly, it's not that far from the truth. Imagine providing your learners with the critical learning or performance support snippet exactly at their moment of need—when they are on the job and performing the task. By providing them with the right content in the right context, you help them perform their tasks more quickly and efficiently and at that moment when they need it the most. Maybe it's not as dramatic as a superhero saving the day, but if you can provide learners with the information and support needed when it is most needed, you're saving the day not only for them but also for the company. You can rest assured that your employees will be enabled to quickly make the best decisions and perform at maximum effectiveness. This is an extremely powerful benefit of mobile learning.

Location- and Context-Specific Content

Ever launched an app on your smartphone, only to have it tell you that you must turn on the location-based services to run the app? Have you seen the message boxes asking if you would like to share your location with others on your social networks, such as Facebook or Twitter? One unique attribute of mobile devices is the ability to use GPS technology and location-based applications. Think for a moment about the power we have in our hands when the technology knows not only where we are physically located but also what we are looking at on the device's screen. The context we can provide to our learners is incredible.

Data Collecting

As we've all heard before, knowledge is power in business, and data is what gives us the intelligence we need to make the best decisions. In mobile learning, there are two primary ways that data collecting is a benefit.

FOR EXAMPLE

Learning About the Night Sky

I always wanted to learn about astronomy, and as a child I could often be found lying in my backyard staring at the night sky. In college, I seriously thought about taking an astronomy class, but dropped the idea when I discovered the course involved a tremendous amount of math, something I avoided like the plague. So my dream of learning about the stars went on the back burner—until I purchased my first iPad and downloaded Star Walk, an augmented reality program that identifies the stars, constellations, and planets. Now, as I gaze toward the heavens, I simply hold my iPad toward the night sky and the program shows me everything I'm looking at, even giving me a tour of that evening's sky. All thanks to mobile technologies and location-based services.

The learner can provide us with data he or she has collected and send that information to someone to review, give feedback, approve, and take further action. For example, a salesperson completes a remote task such as appropriately setting up an end cap (a display of products at the end of an aisle) in a store. When the salesperson believes he has correctly positioned all the products on the shelves, he takes a picture of his work and sends it to his mentor or manager to review and provide feedback. That salesperson isn't even limited to still pictures of his work. He could also use the video camera or the audio recording feature of his mobile device as a way to collect on-site data and provide management with evidence or data on his remote learning activities.

A second way you can see data collecting as a benefit is by tracking your learner's interactions with the device and the web. For example, you could track how she is using the device for learning, which quiz questions she is struggling to answer correctly, and trending how she responds to challenges.

Extending the Learning Process

In a typical instructor-led training course, students will most likely forget about 50 percent of what they have learned as soon as they exit the building. The instructor probably provided too much information at one time, and if the students did not have the opportunity to apply or reflect on what they've learned, they will continue to forget that information.

Mobile learning gives us an opportunity to extend that learning experience by spacing the delivery of content and providing smaller nuggets

TRICKS OF THE TRADE

Bringing It Together

When you get right down to it, in one way or another, all of these learning benefits revolve around increasing the productivity and/or the performance of your employees. Here are some of the ways you can improve productivity and/or employee performance:

- Transform dead time into productive learning time.
- Provide immediate access to critical information that will allow learners to troubleshoot and make educated decisions.
- Improve the application of learning on-the-job content by providing short, targeted, contextual educational content at appropriate moments.
- Increase collaboration among employees such as sharing best practices with user-generated content.

of content. This could be accomplished by blending a mobile solution with an existing program. The students could receive questions asking them to apply what they learned in the course. Based on their answers, you could follow up with immediate feedback and even send them links to new information to help them understand the correct responses.

Challenges of Implementing Mobile Learning

As with deploying any new technology, the road will not be devoid of potholes. In addition to overcoming the myths and misconceptions discussed in Chapter 1, you may face some of the following challenges when you launch your new learning solutions: costs, change management, and integrating data with internal systems. In this section, I provide you not only with the most common challenges but also ways to help you overcome them.

Challenge: Costs

It's common that when executives first think about mobile learning, one of their primary concerns is the cost associated with implementing the solution. Common cost-related concerns include some of the following questions:

- How much will it cost to redevelop existing courses into a mobile format?
- Will we need to purchase new development tools to create mobile learning?
- Do the employees need to be trained in these new tools?
- How much will they spend on integrating the learning tools with other internal systems?
- Must the company buy mobile devices for every employee?

While all of these questions are valid and should be evaluated before jumping into a full-blown mobile learning rollout, they shouldn't be viewed as deal breakers.

Consider the following when you're presented with the challenge of the perceived high cost of mobile learning. In many cases, mobile learning can be more cost effective than traditional learning experiences such as instructor-led training and e-learning. When talking costs, consider how your organization will either save money or make money by leveraging mobile learning.

For example, how much money would you save by reducing the time employees spend in a classroom and increasing the productive time they spend on the job? What if salespeople could close a sale twice as fast if they had critical sales intelligence right in the palms of their hands while interacting with prospects? How much money is saved by the shortened sales cycle? How many additional sales over the course of the year could each salesperson close with the shortened cycle? When you design your mobile solutions to harness the power of increased performance and job productivity, ask yourself, "Do the benefits outweigh the costs?" You will find that they usually do.

Challenge: Change Management

A key challenge that needs to be addressed proactively when implementing a mobile learning solution is the organizational changes that will occur when rolling out a new technology. My first recommendation here is to assign one person ownership of and responsibility for managing the change process. A change such as this requires frequent messages to employees that emphasize the benefits of the rollout to everyone in the organization, the type of support provided during the rollout, and how employees can provide feedback on their experience so the system can be improved. If your company has an incentive program in place, you should explain how this initiative fits into the incentive program.

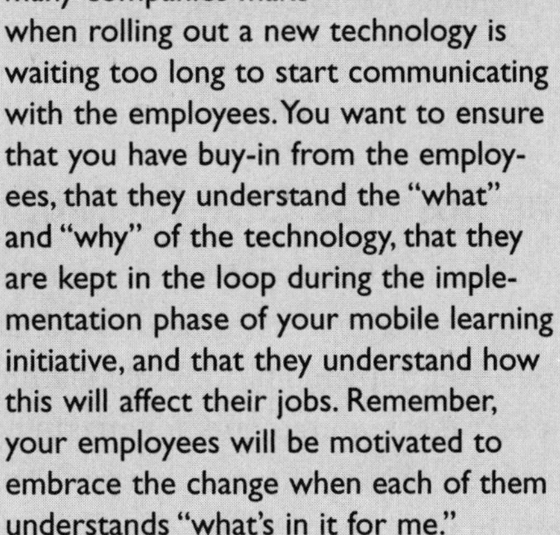

ENSURING SUCCESS

One of the most common mistakes that many companies make when rolling out a new technology is waiting too long to start communicating with the employees. You want to ensure that you have buy-in from the employees, that they understand the "what" and "why" of the technology, that they are kept in the loop during the implementation phase of your mobile learning initiative, and that they understand how this will affect their jobs. Remember, your employees will be motivated to embrace the change when each of them understands "what's in it for me."

Challenge: Integrating Data with Internal Systems

A challenge that managers may face in rolling out a new mobile learning program is the collection of user data and integrating this with legacy enterprise systems. These systems may include a talent management system, a human resources information system, or even a learning management system (LMS). While integrating the data can be a challenge, you can accomplish this using a variety of methods ranging from templates, to single sign-on processes, to application program interfaces (API).

Partner with your IT department to address this challenge and include IT in key conversations regarding the tracking, storing, and sharing of data among various systems. You need to ensure that the IT department understands the nuances of the existing systems into which you want to integrate new data. Often, the IT group will recommend that you include a technical partner from a particular business unit, such as Human Resources or Learning and Development, who may have a better functional understanding of the systems or in some cases may be the owner of the technology into which you are integrating the new data.

KEY TERMS

Learning management system (LMS) An enterprise-level software program designed to manage your corporate learning initiatives, including administration, curriculum access, tracking, and reporting.

Application program interface (API) A specification written as a means of interfacing software programs, web services, or modules to allow communication among systems.

The Business Case for Mobile Learning

Now that you have a good understanding of the primary business drivers, the potential benefits to be derived, and the challenges you might face when you implement a mobile learning solution, let's take a high-level look at the requirements for making your business case. As you work through the later chapters, we go through each of the processes step by step. In fact, when you finish this book, you should be able to write not only a solid business case that will stand up to the scrutiny that these type of presentations typically receive, but also have a solid plan in place to ensure successful implementation of a mobile learning initiative.

When I started consulting with clients on the design and development of custom learning solutions, I struggled with documenting the business case. It always seemed like a daunting task. Then one day the lightbulb went on, and it didn't seem so daunting anymore. Developing a business case was an opportunity for me to explain my client's current state in the context of a specific business problem and compare that with the desired future state, including the primary goals and success criteria. We could explain the vision for the solution, including the benefits, success metrics, challenges in moving forward, and the budget and time needed. I now think of a business case as the opportunity to gain perspective on the current state and to share my vision for the desired future state.

Questions to Answer When Writing Your Business Case

TOOLS

When writing your business case for adoption of a mobile learning system, you must be able to answer the following questions:

- What is the business problem that you're solving?
- What is the organization's current state, and what will be the organizational impact of adopting the mobile learning initiative?
- If you address the problem, what would be the future state of the organization?
- What are the success criteria and metrics for the future state?
- How does your solution solve the problem, and how will you measure success?
- What are the core drivers, benefits, and challenges of mobile learning specific to your business problem?
- What are the solution's short-term and long-term goals?
- What additional resources are required to implement the solution?
- What return on investment do you anticipate by implementing this solution?

Manager's Checklist for Chapter 2

☑ Mobile learning provides a way to support our modern workforce that includes a highly mobile group of individuals known as the Millennial generation.

- ☑ Mobile learning provides a way to improve employees' on-the-job performance.
- ☑ Mobile learning improves the overall effectiveness and impact our training solutions have on the organization.
- ☑ The benefits of mobile learning include anytime/anywhere access to learning content, just-in-time training opportunities as well as performance support, provision of location- and context-specific content to employees, increased data collection opportunities, and extension of the learning process to increase the impact on the organization.
- ☑ Challenges that you may encounter when deploying mobile learning include up-front development costs, integrating data with legacy systems, and keeping your employees up to speed on the changes taking place within the organization.

A New Way to Think About Training

Learning and innovation go hand in hand. The arrogance of success is to think that what you did yesterday will be sufficient for tomorrow.
—William Pollard

William Pollard's quote is especially meaningful in regard to training solutions and mobile learning. Whenever a new technology arises that will help us train our employees, our first instinct is to use that technology in the same way that we used the last technology. Let's look at a few examples. In the 1980s and 1990s, many organizations tinkered with video as a way to replace their instructor-led training. In the late 1990s and into the early 2000s, companies looked at e-learning as a way to replace their instructor-led training. In most situations, neither of these approaches worked well.

Neither delivery method provided the core element of interaction that would help most individuals learn, retain, and apply what they'd learned in the context of their jobs. After rolling out these new technologies, many organizations learned that the best approach was to blend them into their existing approaches to harness each technology's abilities. For example, video is often used today as a different way to demonstrate a complicated process. Its true power is to show the learners how to perform a task or to illustrate the importance of visual cues.

Have we learned from our mistakes? Are we looking for new ways to deliver training and support our workforce, or are we approaching mobile

Instructor-led training (ILT) Formal training that takes place in a fixed, predetermined environment in which both the students and facilitator/instructor are in the same setting. ILT can include classroom training, workshops, and lectures.

e-Learning A formal, self-paced training program where the learners interact with a computer program and access the training program via the Internet. e-learning does not include real-time interaction with a facilitator/instructor or with other students. e-learning is also referred to as web-based training or online training.

learning as a means of consuming our e-learning programs? Are we looking at the unique attributes of mobile learning? Are we thinking outside the box about how we can provide learning opportunities to our staff, or are we attempting to use this new technology to deliver the kind of training that we now offer?

In this chapter, we address whether creating mobile versions of your e-learning solutions truly is the best use of this new technology. We look at when learning really takes place in the workforce and discuss how you can harness the true power of mobile learning to your training strategy. First let's start with a discussion of whether you should convert your existing e-learning programs to a mobile delivery format.

Should You Create Mobile Versions of Your e-Learning Courses?

When I speak at conferences, I routinely arrive in the room at least 45 minutes early. I like to make sure I have time to get the equipment running and get comfortable in the setting. This also gives me a chance to speak one-on-one with some of the early arrivals and confirm that my session material is on track with the audience. One morning, as I was getting ready to deliver an overview session on the power of mobile learning, I strolled into the room and was surprised to see one of the attendees had already arrived. After some basic chitchat, I began setting up the equipment, and he asked, "I know this is probably a silly question and I'm sure you will cover it in your session, but what exactly is mobile learning? The only thing I can think of is it's like the continuing professional education courses I take online, but now I'm supposed to take the courses on my phone." I immediately stopped what I was doing and replied, "Gosh, I hope not. That would be a terrible experience." He quickly agreed. While I assured him that we would discuss this in depth during the session, I added that mobile learning is *not* taking your e-learning courses on a mobile device.

When people first hear the term *mobile learning* or start thinking about how to leverage mobile technologies in learning solutions, it is common that they begin looking at it from the perspective of a new way to deliver the same content. With this in mind should you convert your existing e-learning courses to a mobile format? Is that mobile learning?

I am a believer that under most circumstances you should not convert your existing e-learning libraries and courses to a mobile delivery format. The thought of taking content that was specifically designed for one medium (i.e., the desktop computer) and converting it to another form like mobile phones makes me cringe. Why is our knee-jerk reaction to use new technologies to accomplish the same old goals?

CONVERTING LEARNING CONTENT

CAUTION

A common mistake managers make is to believe they can effectively convert learning materials from one format to another. Consider this: Would you convert a face-to-face workshop session into an audio recording that your learners could listen to instead of attending the session? Why not? After all, it's the same content and you would definitely save the costs of taking your employees away from their job and bringing them together in one room. Sure, they can listen to the audio files at their convenience, but is it the right choice? No, definitely not.

In this example, we have completely modified the experience and affected the potential outcomes that we want the learner to achieve. The learner interactivity, engagement, and context are completely different in the two mediums. There is a huge benefit to the interactions the learner has with the other participants, and you lose the opportunity for your learners to critically think through and participate with the material. You are taking an active learning experience and converting it to a passive experience. Chances are good that the learner will not master the learning objectives or be able to apply what he or she learned on the job.

Remember: learning solutions should be designed specifically for the delivery method. The same principle is true with mobile learning. Be careful not to fall into the trap of thinking that training materials can be converted to another delivery method and still be effective.

Think about some of the e-learning courses you provide to your employees. Can you imagine taking them on the tiny screen of a smartphone? While there are many vendors in the marketplace that focus on converting PowerPoint files and existing e-learning courses to a mobile delivery format, in most cases, is not the best use of your time and money. Then what is mobile learning good for, if not to create a mobile version of an e-learning course experience? Mobile learning provides us with an opportunity to rethink how we deliver training opportunities to our employees. Before we get into the details of what mobile learning truly is and how you can harness that power in your organization, let's look at when learning really takes place in the workplace.

When Does Learning Really Take Place in the Workplace?

In this section, we look at two key points. First, we look at five main ways

Questions to Consider Before You Convert Content to a Mobile Delivery Format

Here are a few questions you should ask yourself before jumping into converting your e-learning courses for mobile devices. First, what is the intent of the e-learning curriculum that you want to convert? For example, at the end of the course do you expect a behavior change to take place? If the answer is yes, then converting your e-learning to a mobile delivery format is a bad idea. If your intent is that the learner will gain a basic understanding or awareness of the content, then converting your content to an alternate delivery method might be okay. However, you must consider the type of device employees will use to access the content.

Do you anticipate that your learners will take the courses on a tablet or a smartphone? If smartphone is your answer, then ask yourself this question: Can the same level of user experience be achieved on the mobile version that the computer-based e-learning version offered? In most cases, the answer to this question is no. However, if your answer was "tablet," then you *may* be successful in converting your e-learning to a new format and re-create the learning experience. I would, however, strongly recommend that you not refer to this experience as mobile learning, but rather e-learning delivered on a mobile device. First, mobile learning takes into account that both the learner and the device are mobile. Second, mobile learning is about providing a unique experience based on mobility. Mobile learning is not about a new form factor for delivering content.

organizations offer training opportunities to their employees. These are by no means all of the training delivery options available, but the most common ones. Then we discuss the circumstances under which employees are motivated to learn on the job.

The Training Opportunities Companies Provide Their Employees

The first training method is *instructor-led training*. For decades, organizations have been taking their employees away from their jobs and sending them elsewhere to learn new skills in instructor-led training courses. Some of these experiences are in a public setting with a predefined, generic curriculum, and other experiences are in a private company-only setting with a customized agenda to meet the specific needs of a company or group of individuals.

> **PUTTING THE DELIVERY METHOD FIRST**
>
> CAUTION
>
> A common mistake managers make is looking at training only from the perspective of the types of training offerings that are at their disposal. They assume that those delivery methods must be the perfect fit for their problem. There is no silver bullet in training, and each method has its strengths and weaknesses. Make sure the delivery method or the blended solution will solve the business problem you are trying to address.

If an employee needs to improve his or her skills in a particular area—let's say learning to write complex formulas using Microsoft Excel—it would be common for the employees to take a full-day Excel course. During that full-day course, they are exposed to a tremendous number of topics that may have nothing to do with writing formulas and may have no impact on their duties.

Regarding the content on writing formulas, they may learn the basics of how to write a formula and practice using several formulas, but chances are they will not learn all the formulas they need to know for their job. Nor will they have a chance to practice the formulas they do learn in the context of their job duties. In instructor-led classes, the students typically forget about 50 percent of everything they were taught in the session as soon as they walk out of the door. That percentage increases if they do not have an opportunity to immediately apply what they've learned in the context of their specific duties. In addition to forgetting most of what they learned, the responsibility is on the learner, in most cases, to convert the basic content to their specific application.

Another common delivery method for corporate training is through *e-learning courses*. With e-learning, we do not ask our employees to leave their job, but rather, to fit the learning into their business day and to typically take the training at their desk. Some organizations allow their employees to take the e-learning courses from their homes or places outside of the work environment. This easy, anytime access to learning content sounds great, but the learners can be bombarded by the distractions of their job and may not give the training experience their full attention. Employees often try to multitask while taking the training. The phone may ring with a time-sensitive call, they may feel compelled to answer e-mails, or someone may enter their office or cubicle in need of an answer to a pressing problem.

Even taking training outside of the office is not a distraction-free zone. On the upside, depending on how the e-learning system was designed, the employee may have an opportunity to choose the specific content he or she needs to learn for the job and not have to sit through all the lessons, lessening the amount of information that they forget. e-Learning also provides an easy way for employees to go back and refresh their memory on a particular content area in which they are struggling. However, they do lose the opportunity to ask an instructor questions and learn from other employees' questions or comments. And they must assume the responsibility of converting the content they learned to their specific application.

KEY TERM

Distance learning A formal training program where the students and instructor use the Internet to log in to a virtual classroom. Distance learning occurs at a predetermined time, but the attendees may participate in the sessions from different locations. Webinars and teleconferencing sessions are examples of distance learning.

Distance learning is another delivery method. Distance learning provides the benefit of interaction between the instructor and the students, but still tends to be a passive experience for the learner. As with e-learning, students may be distracted by workplace activities and attempt to multitask during the session. Distance learning is a lower-cost delivery option than instructor-led training as there are no facility or travel expenses. While this training does allow for interaction between the facilitator and the students, typically the communication is one-way, from the instructor to the students. Some distance learning programs are recorded, which allows the employees to review the content. However, it is a challenge to find the specific content block that the employee needs to watch and so it's rarely used in this fashion. More frequently, employees who were unable to attend the session at the predetermined time watch the recording later at their convenience.

Mentoring is another way companies provide training to their employees. In this model, a more experienced employee (the mentor) takes the responsibility to show another employee the ropes and the mentor is available for questions. This experience provides the context that is typically needed for a learner to transfer the knowledge to the job. Although mentors are typically experts at a specific job, they are usually not experts at providing a training experience and may have difficulty in transferring their knowledge to someone else. A mentor may not always be available to the employee, or the employee may feel uncomfortable admitting his or her lack of knowledge to the mentor.

The final training method we discuss is using tools such as *job aids*. These are documents or files that typically outline the steps needed to complete a process or task but tend to lack or overlook the underlying context of the material and the "what's in it for me (the employee)." Job aids are a great way to ensure the learner can apply the knowledge that he or she has learned in other training programs to the job or can even successfully perform a task in which he or she has never received training.

> **KEY TERM**
>
> **Job aid** Any source of information that provides guidance on accomplishing a particular task. Job aids break down a task into step-by-step instructions. The intent of a job aid is to provide support while an individual is actually performing the task.

These are by no means all the ways that companies provide training opportunities, but they tend to be the most common. All have their advantages and their pitfalls, and none of them are the silver bullet for

KEY TERM

Blended learning An approach to delivering learning interventions that weaves together multiple delivery methods in a single training solution. For example, requiring employees to take an e-learning course to establish a baseline understanding on some content prior to attending an instructor-led event where they are provided with enhanced content and job aids.

FOR EXAMPLE

Blending Delivery Methods to Optimize Performance

A company was rolling out a new software product that would drastically change the way in which the employees performed their job. The training solution included e-learning, instructor-led training, mentoring, and on-the-job support aids. The first step, the e-learning course, started with a short video featuring a senior executive who explained the benefits to the employees of transitioning to the new enterprise software package. The executive acknowledged the potential frustration the employees might face in learning new processes to accomplish their jobs. Then the employees actively participated in a series of basic software simulations followed by an online assessment. Next, they attended an instructor-led course with others from their department.

Prior to the course, the instructors analyzed the assessments and identified areas where the groups were struggling, which allowed the instructors to customize each class. In the course, the students not only learned advanced software functions, they also asked questions on how the new software and processes would change their jobs. During the classes, the instructors identified power users who would provide on-the-job mentoring to other employees in their department. The final step in the process was to use job aids while learners performed their job until the new skills became a habit.

This blended approach leveraged the strengths of each delivery method and provided the employees with the knowledge and support they needed to successfully adapt to the new software and processes.

training our employees. Most organizations create a blended learning strategy to maximize the impact of the training for the employees.

What Motivates Employees to Learn on the Job?

In the previous section, we established five common ways that companies provide training: instructor-led training, e-learning, distance learning, on-the-job mentoring, and tools such as job aids. We also touched on how one company successfully blended these in a planned strategy for success. Most of us do not wake up in the morning and say to ourselves, "Gosh, I can't wait to learn today." So what drives our employees to learn on the job? Whenever I explain this topic, I like to use Bob Mosher and Conrad Gottfredson's model of the Five Moments of Need. These are the moments when our employees need to learn and where we as managers and organizations need to provide support to help them perform their jobs. This support comes in the form of both formal learning and informal learning events.

Formal learning Any learning where there are predefined learning objectives and desired outcomes. The instructor, training department, or company sets the learning objectives and the appropriate content to be covered to achieve those objectives. A few examples include instructor-led training, workshops, seminars, and e-learning.

Informal learning Any learning where the individual learner controls what content he or she is to learn and the associated objectives. There is no predefined curriculum. A few examples include performance support, social networking, and expert mentors.

Let's look at how Mosher and Gottfredson explained the Five Moments of Need in an article in *Learning Solutions* magazine, "Are You Meeting All Five Moments of Learning Need?" Quoting from the article, the five moments of need are:

1. When employees are learning how to do something for the first time (new)
2. When employees are expanding the breadth and depth of what they have learned (more)
3. When they need to act upon what they have learned, which includes planning what they will do, remembering what they may have forgotten, or adapting their performance to a unique situation (apply)
4. When problems arise, or things break or don't work the way they were intended (solve)
5. When people need to learn a new way of doing something, which requires them to change skills that are deeply ingrained in their performance practices (change).

In breaking down these five moments, the first two—when learning something for the first time and when students need to learn more—represent opportunities where formal learning makes the most sense in providing support to our employees. These formal methods work best when new information needs to be learned or when we need to learn more about a particular topic. The last three moments focus on applying knowledge, solving problems, and navigating change—all within the context of students' jobs. Informal learning opportunities have been found to best support these situations.

If we take a look at training and development budgets, most of an organization's investment is made in formal learning opportunities; however, when on the job, usually what our employees need falls along the lines of performance support. So as managers, how can we provide our employees with a well-rounded strategy that addresses all of their moments of need? Many gurus in the learning and development field strongly believe that performance support is the "sweet spot" for mobile learning. Now that we understand *when* our employees need support and *what* motivates them to learn in the workplace, let's take a look at *how* mobile learning can support those needs.

Harnessing the True Power of Mobile Learning

In this chapter, we have discussed how mobile learning is not about recreating your e-learning courses to fit on a smaller screen. However, if it's not about an alternate delivery method for your e-learning, then what is mobile learning? Briefly, mobile learning is about leveraging the attributes that are unique to the employee's mobile device to create an experience that will increase your employees' retention and ultimately improve their performance. We can think of the true power of mobile learning as a way to augment the formal learning process and as a key ingredient in providing our employees with the performance support they need to accomplish their tasks. In Chapter 4, we break these concepts down to specific types of mobile learning, but in this chapter we focus on the big picture.

Augmenting Formal Training

Formal training accounts for approximately 20 percent of the total knowledge that employees need on the job, yet most corporations invest the preponderance of their time, resources, and training and develop-

> **RETHINKING WHAT MOBILE MEANS**
>
> CAUTION
>
> A frequent mistake that managers make when designing mobile solutions is to think about mobile learning only as a device that employees will use to access content. Then the managers look for ways to fit their existing programs into that technology.
>
> Successful mobile learning initiatives instead focus on the following points: (1) The learner is mobile. How can you create a unique experience to support him or her using mobile technologies? Sure, the learner is using a mobile device, but you really need to think about the context—where the person is physically and what he or she needs at that specific moment. (2) Mobile learning is about leveraging the capabilities unique to the devices.
>
> Mobile learning is not about looking at the devices as simply a smaller portable version of the learners' desktop computers. Think about how you can leverage the camera, video recorder, microphone, sensors, location-based services, touch screen, text, and other unique features of these devices.

ment budget on the design, development, and implementation of formal training events. One of the real powers of mobile learning is its ability to augment the formal training process and to increase content retention and application. A few ways to accomplish this are:

- Spacing out your learning content instead of providing all the learning content during one event;
- Providing your learners with reinforcement opportunities after the formal event;
- Providing learners with opportunities to reflect on and apply key concepts;
- Using games to increase learner engagement and providing a safe environment to practice new skills; and
- Establishing communities of practice to share best practices and learn from other employees.

> **Communities of Practice**
>
> 
>
>
> A group of people from the same profession who share experiences and insights, such as tips and tricks and best practices, and provide support for each other specific to their profession.

One of mobile learning's key benefits is the ability to spread out learning content over time to increase the learner's retention of the material and increase the program's effectiveness. This theory, called "the spacing effect" (an approach where information is easier to understand if repetitions are spaced out over time), is supported by research and was developed in 1885 by Hermann Ebbinghaus. Prior to development of mobile technologies, in most cases to leverage his theory in practice in the workplace was cost prohibitive. However, thanks to mobile technologies, organizations can now incorporate the spacing effect into their training programs.

One of the unique attributes of mobile phones and now even some tablets is the ability to program timed SMS text messages. This technology has a tremendous amount of power in mobile learning solutions by providing your learners with timed reminders and learning reinforcement in the palm of their hand. For example: you can preprogram messages to be sent at a future time, including short, text-based content, video links, and even assessment questions.

FOR EXAMPLE

Mobile Learning Provides New Parents with Timed, Critical Care Learning

Imagine you're a first-time parent who doesn't know what to expect. In the past, many first-time parents purchased books, such as *What to Expect When You're Expecting*. Now Johnson & Johnson has looked to mobile technologies to provide spaced learning to new parents to help them navigate these unknown waters. Text4Baby is a free service that educates new parents on what to expect not only throughout their pregnancy, but also through the first year of the baby's life. Expectant parents enter the baby's anticipated due date into the program, and Text4Baby sends them educational text messages timed to coincide with how far along the pregnancy is or how old the baby is. The educational messages focus on critical health concerns and tips such as prenatal care, breastfeeding, immunizations, labor and delivery, car seat safety, injury prevention, nutrition, and much more.

Another way in which you can apply the spacing effect and augment your formal learning programs is by sending out pre- and postcourse work. Imagine the benefits of sending out short videos prior to a formal event to get people up to speed on the content to be covered and also as a way to gain buy-in into the importance of the content to their jobs.

Providing Performance Support

As cited earlier, only about 20 percent of learning on the job takes place

QUESTIONS ON AUGMENTING YOUR LEARNING PROGRAMS

To get you thinking about how you can benefit from and augment your learning programs with mobile learning, consider these questions:

1. How can you provide snippets of content to reinforce key content from other learning events such as instructor-led training or e-learning?
2. How can you provide opportunities for your learners to reflect on key content blocks?
3. What type of timed events can you use to trigger content delivery to your learners?
4. What content are your employees struggling to apply in the workplace?

in a formal setting, which leaves 80 percent of learning to occur through informal channels such as performance support, learning from peers, and working alone through problems on the job. With the increased demands of most jobs, the amount of complex information we are required to remember is increasing.

One area where people are poor at remembering is lists of information, but many organizations rely on their employees to remember just such lists. Fortunately, mobile devices are excellent for that function. Performance support can provide your employees with the tools to access this kind of information so they can perform these new, complex tasks and maximize their capabilities. You can create and send great performance support applications to your employees such as job aids, coaching, and mentoring so they can solve their problems while they are in the field.

One way that mobile learning supports performance is by providing access to mentors or experts to walk an employee through a process in his or her moment of need. Imagine for a moment being a doctor in a remote location and encountering an emergency. Let's say you *must* amputate a limb immediately, and you are unfamiliar with the exact steps of the surgery. What do you do? How can you learn exactly how to perform that surgery in your patient's critical moment of need? Are you thinking the answer is mobile learning? You're right. By contacting an expert on the procedure, that expert can text the field doctor the step-by-step procedure. Sound far-fetched? Amazingly, there are quite a few stories of doctors doing just this and, yes, saving lives using perform-

FOR EXAMPLE

RED CROSS PROVIDES SUPPORT DURING A CRISIS

By developing a series of mobile applications, the Red Cross has modernized its approach to providing assistance to people in the midst of a crisis. Each app is specific to a particular crisis, such as a hurricane. By using the app, people in the affected area can prepare themselves and their home before the storm hits, communicate with family and friends via social networks about their safety, and find help after the event has passed. Some of the apps' support functions include the location of nearby shelters, checklists for creating an emergency plan, and even a strobe light and an audible alarm to help rescuers find them if they are trapped. The app content is loaded directly onto people's phones so access to critical information is available without a mobile connection.

ance support delivered by mobile technology.

In this chapter we have laid the foundation for two main ways we can effectively use mobile learning within our organizations. First, as a way to augment our existing learning programs to increase employee effectiveness, and second, as a vehicle to provide in-the-moment performance support. In the next chapter we dive into specific ways in which we can use mobile technologies to accomplish these goals.

Manager's Checklist for Chapter 3

- ☑ Converting your existing e-learning courses into a mobile delivery format to play on your employees' smartphones is a poor idea. However, if you do convert your e-learning courses to play on tablets, do not classify the experience as mobile learning but rather as an alternate delivery method for your e-learning. Think of it as mobile e-learning.
- ☑ Mobile learning is about the learners being mobile, as well as learners using mobile devices. You should provide unique learning experiences with this in mind.
- ☑ Companies need to provide learning solutions to support all the moments of need that employees encounter on the job. This includes both formal events and informal performance support solutions.

- ☑ Mobile learning is a way to increase retention and apply learning content by augmenting formal learning events.
- ☑ Mobile learning is about creating a learning experience that ultimately improves your employees' performance on the job by leveraging the unique attributes of mobile devices.

Types of Mobile Learning Experiences

Now that we've reviewed the big picture of mobile learning, we delve into the details of the types of mobile learning experiences you can provide your employees. In this chapter, we discuss the importance of providing an experience that fits your employees' needs and the four types of experiences for which your company can use mobile learning. We discuss how you can use mobile learning to provide information to your employees, how you can gather data about them, and how you can increase communication and collaboration. Last, we look at some cutting-edge and innovative ways to leverage mobile learning.

It's Not About the Technology, It's About the Experience

When company managers begin to think about using mobile learning, often their first thoughts go to the technology element. They investigate the types of devices that they will use and fret that their employees do not all have the same device. Managers worry about the costs associated with standardizing their employees' devices and wonder how soon those devices will be out of date. Managers focus on the differences among the technologies, such as screen size, and wonder how they could possibly develop a solution that delivers a consistent experience regardless of learners' devices. Managers worry that bandwidth will be an issue and how they will track data within their existing corporate-wide systems. While all these concerns are valid, a smart manager does not start with the technology, but rather, focuses on the user experience. The employees and their needs while they are mobile are the key issues.

Begin by thinking about how your employees already use their mobile devices. Do they spend an hour or more on the device at a time? Most of the time, the answer is no. Instead, they use it for short bursts of activity. Five minutes here, ten minutes there to perform a specific task. For example, if they have a question, they use the Internet with their mobile web browser and google the question or go to YouTube to view a short video.

Other frequent uses of mobile devices are to stay connected with others and to communicate with others. Employees make calls or, more frequently, especially with younger generations, they text. We stay up-to-date with family and friends using our social networks. For many of us, social networks such as Facebook or Twitter let us reconnect with people from our past, whom we may not have spoken to for eons. We now share important news, family photos, or funny comments with our friends and family at the moment they happen, not weeks later. No longer do we need to wait to see the holiday photos of the kids visiting Santa or the pictures from vacation. Instead, we visit their Facebook page.

We also use social networks as a resource to ask questions. People will often reach out to their online networks for advice and guidance with personal as well as work-related problems. I recently saw on my LinkedIn network a business connection who was interested in using gamification as part of her company's training solutions. So she sent a status update to her network asking anyone who had information about the topic to please contact her. In addition, I often see my Facebook friends asking how to accomplish work tasks such as the best way to create a data entry form in a particular software product.

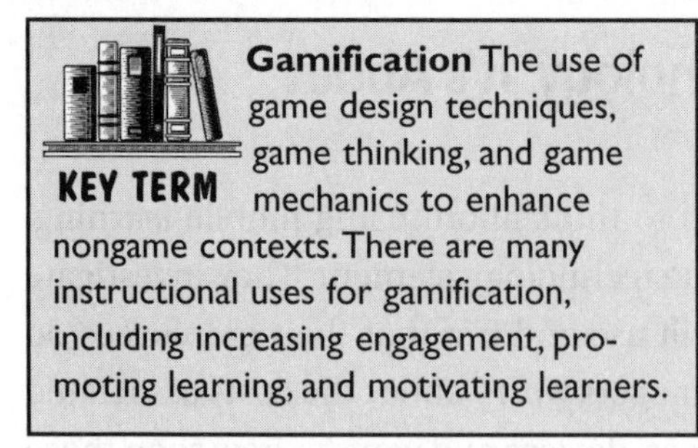

KEY TERM

Gamification The use of game design techniques, game thinking, and game mechanics to enhance nongame contexts. There are many instructional uses for gamification, including increasing engagement, promoting learning, and motivating learners.

When thinking about mobile learning in your organization, keep in mind how your employees already use the technology; then think about how you can provide support that fits the usage. The following sections of this chapter break down how you can leverage mobile learning as a way to address those needs in your organization.

Using Mobile Learning to Provide Learners with Information

When managers typically think of providing training to their employees, their first thought is to provide them with information. This is one of the most common ways in which we can create a mobile learning experience: allowing our employees to retrieve the information they need at the moment they need it.

RETHINKING HOW MUCH TIME IT TAKES TO LEARN

SMART MANAGING

Most managers think of learning taking place over a long stretch of time. We send our employees to full-day, instructor-led courses or assign them to complete a two-hour e-learning course or have them attend a one-hour webinar. Mobile learning requires us to rethink the amount of time it takes for our employees to learn what they need to accomplish.

When providing information to your employees through mobile learning, your content should be short snippets or bite-sized chunks versus long courses. Think about content length in terms of minutes, not hours. To help you reprogram your mindset, begin by anticipating the questions and problems your employees are faced with at their moment of need on the job, then provide an easy way to find the answers instead of requiring them to memorize everything ahead of time.

Podcasts and RSS Feeds

Podcasting provides a way for you to present information to your employees in a fun and engaging way, allowing them to learn on the go and at their convenience. We discussed in Chapter 1 that a podcast is a multimedia file (audio or video) that you access from the Internet. But what is the experience like?

Let's discuss two of the most common podcast formats. The first is a talk show approach, where the show's host brings in a subject matter

FOR EXAMPLE

Using Podcasts to Learn How Executives Use iPads in Their Businesses

One example of using a podcast to learn about a topic is Apple's Execs Talk iPad podcast series. This video podcast series provides short interviews (five minutes or less) with executives on how companies, such as Boston Scientific, Saleforce.com, Coldwell Banker, Greentec, and NBC Universal, use this technology. The series format is that of a talk show, and each executive participates in multiple interviews. Each interview (podcast) focuses on a specific management subject, how each executive deploys the technology, or how he or she measures success using such technology.

expert and interviews him or her. Talk show formats are typically no more than 15 minutes long, often as short as 5minutes. The second common format features an expert host who shares his or her insights or tips on a topic. These podcasts are typically only a few minutes long.

Ideally, in your own situation using either format, you are producing a series of podcasts at regular intervals that focus on a particular subject area such as plant safety, sales, or customer service.

Regardless of the format you choose for your podcasts, most likely you'll host your podcast series on a website, and your employees will subscribe to an RSS feed using an RSS reader program. An *RSS reader* aggregates all of each employee's selected feeds into one simple-to-access location. *Aggregation* saves employees the time of having to visit individ-

CAUTION

Experts Need Help, Too

Just because someone is an expert on a topic, that does not necessarily mean he or she knows how to break down a topic in a way that others can understand or that he or she is a good speaker. Not every expert will need assistance, but I recommend that you designate an experienced coach or instructor to assess the expert's skill level in presenting content.

The coach/instructor can help the expert script out what to say so he or she provides just the right amount of information and presents it in a way that your employees find easy to understand. The coach/instructor can also provide the expert with tips on good presentation techniques, as well as ensure he or she engages the audience. This may sound time consuming, but it's really not. Just make sure your coach/instructor is an expert in creating engaging and effective podcasts.

ual websites and manually search for the content. Once a user has subscribed to an RSS feed, the software checks each site for new content and downloads the content at predetermined intervals. The user can unsubscribe from the feed at any time.

> KEY TERM
>
> **RSS feed** A delivery mechanism (RSS stands for *rich site summary*, also referred to as *really simple syndication*) that syndicates or distributes online media such as new articles, blogs, videos, and podcasts. The benefit of an RRS feed is that, once you subscribe to the service, you no longer have to visit the original source. The RSS feed ensures that you have all the latest media files you're interested in.

Videos

Research shows that we remember about 70 percent of what we can see and hear. With this in mind, you can see that video can be a powerful medium for mobile learning solutions. You can create a video in many ways. They can be computer generated using computer software, professionally recorded, or even recorded using the video application on your smartphone. In the past, there was a belief that videos had to be high fidelity and professional production quality with high-end special effects to be acceptable for training. With the vast availability of videos on the Internet today, that notion is long gone; people have learned that the most important criteria is good content.

> MISTAKE PROOFING
>
> **Two Potential Video Problems**
>
> Two common mistakes with videos: they are too long to hold employees' attention and they are filled with fluff. Remember, your employees are using their mobile devices in short bursts, so why would we think they would take an hour or even 30 minutes out of their day to watch a video that is not for entertainment purposes? In addition to holding their attention, we need to provide easy access to the exact content they need precisely when they need to have it. A long video doesn't allow for that. Sure, employees can fast-forward through the video, but most people won't bother to use trial and error to find the one segment they need. Instead, they will give up looking.
>
> With these two points in mind, try to keep your videos to five minutes or less. Personally, I shoot for under two minutes whenever possible. Also, give your videos descriptive names so it is clear to your employees what the video covers and tag them so they can be found quickly.

Here are a few ideas for ways you can use video.

- Show employees how to perform a complex task.
- Communicate changes within the organization.
- Show how to use a product.
- Show how to perform a complex task.
- Describe how to tackle difficult conversations or negotiations.

e-Books

Step back for a moment to when you were in school. Remember how heavy your book bag or backpack was from all the books that we had to lug around? Luckily for students today, they can carry around a virtual library on their mobile device. They can highlight text, take notes, view others' comments, and search the book for a specific phrase. Many of your employees also use e-books for their personal reading, entertainment, and professional development. If your organization has large manuals and books, consider providing them as an e-book. One advantage you may not have considered with e-books over printed copies, in addition to saving trees and saving money on printing, is that when updates are made to the book, they can be automatically uploaded to the users' devices and replace the old version.

KEY TERM

e-Book A digital format of a book-length publication that is downloaded from the Internet and read on a computer or mobile device. e-Books are most often read using an e-reader such as the Kindle or a tablet such as the iPad. Smartphones are also an option, but the small screen may be too cumbersome for employees to read a large amount of text.

CAUTION

UPDATING E-BOOKS

A great feature of e-books is how you can quickly and cost effectively update and distribute new books to your employees. A downside, though, is that when a user downloads an updated version of an e-book, he or she loses any notes or highlighting in the old version. To avoid this frustration, I recommend sending an e-mail to your staff prior to sending the updated e-book so they can export their notes and highlighting before they download the latest version of the e-book.

QR Codes

QR codes are popping up everywhere. They can be a great way to direct your employees (and even customers) to training content. As described in Chapter 1, you need both a camera and a QR reader on your smartphone to scan the code. You can download a QR reader from your smartphone's app store, and many readers are free. Common uses for QR codes include sharing text-based content, sending SMS messages, connecting to websites, and directing viewers to videos.

Sleep Number Educates Customers with QR Codes and Videos

FOR EXAMPLE

On a recent visit to the local mall, I passed a Sleep Number bed store and noticed a large QR code displayed on their storefront window. I scanned the code and was directed to a short video showing Sleep Number's new product, a fully adjustable pillow called AirFit. The video demonstrated how to easily adjust the pillow's firmness and explained the pillow's unique features. After seeing how you could finally have a perfect night's sleep thanks to this revolutionary pillow, you can either go into the store to test one or click on a link in the video to shop online.

Job Aids

It's common for companies to provide job aids to help new employees apply the content they learned in a formal training program when they are on the job. Historically, companies provided job aids as part of the course manual or as a separate piece of paper. Providing content to your employees with a job aid on their mobile device is an easy way to support them on the go. If they need help working through a task or a process, they simply open the job aid on their phone and read the step-by-step procedure right there.

Reminders

We all need a little extra help in our lives. As managers, I'm sure most of us regularly use some form of electronic reminders to notify us of meetings, to refresh our memory on the agenda, or to refresh our memories on work as we prepare for an upcoming meeting. So why do we not use the same concept for our employees with regard to training? Let's look at how we can support our employees with reminders.

We often ask our employees to complete some type of preparatory work before attending a class. Maybe we want them to watch a video,

FILE FORMATS FOR PUBLISHING YOUR CONTENT

As a manger involved in a mobile learning project, you want to ensure that the content remains accessible on the devices as technology changes. To that end, make sure your Learning and Development team or vendor uses standard file format types that are likely to withstand the test of time. The following table is a list of recommended file formats based on the type of content.

Content Type	File Format
Podcasts	MP3
Videos	MP4
e-Books	ePub
Job aids and reference guides	pdf
Text and images	HTML

read an article, or complete an assessment. Via text message, we can send participants a short reminder of the prework and include links to the assignments. The text messages can be timed to be sent at preset times or, if the prework is tracked online, such as an assessment, we can send out the reminder only if they have not completed the work.

Another way you can use reminders is for postevent support. You can remind the employees of key content covered during the session or provide them with content to help them apply what they learned in the session.

Mobile Applications

Using existing mobile apps or investing in custom apps for learning is another way that we as managers can provide our employees with information. First let's talk about leveraging existing applications. Earlier we mentioned creating e-books. One of the mobile apps your employees could use for their e-reader is the Kindle app. Your employees could use the TED app to access the TED Talk database and watch or listen to presentations on a variety of topics given by well-known experts. Or, your sales force could use the MapQuest 4 Mobile app to get turn-by-turn directions to their sales calls.

KEY TERM

Mobile application A software program (often called a mobile app) developed to run on a mobile device and to perform a specific task. A few examples of mobile apps are the calendar, calculator, address book, notes, games, language translators, web browsers, and mapping programs.

Existing apps aren't always an option, and that's when you need to invest in developing a custom app. They can be developed for any use you can imagine, from providing performance support, training content, and games to teaching your employees how to perform critical workplace skills.

Custom Apps Not Always Necessary

When thinking about using a mobile app, our first thoughts go to developing a custom app specific to our company. But remember that there are thousands of available apps, and their numbers are growing daily. Before you invest in developing a custom app, make sure there is not an existing app that will meet your needs.

NFL Teams Use Mobile App

FOR EXAMPLE

Many of the NFL teams have traded in their 500+ page printed playbooks for iPads and apps. One such app is PlayerLync, which has some of the following features. When the coaches or quarterbacks design or modify a play, they can immediately push it out to all the players. The app goes well beyond the capabilities of a traditional playbook. It has also changed the way players view practice and game footage. Videos are uploaded to the app, giving the players almost immediate access to the footage. They can even review the footage on their plane ride home from a game. Whenever new plays or videos are uploaded to the app, the players receive alerts. Coaches can even see how much time players have spent studying plays and watching the footage.

Do you wonder about security concerns? Don't. The app has multiple levels of security built into the program. For example, if a player loses his iPad or leaves the team, the tablet can be remotely wiped clean.

There are two types of mobile apps: native apps and web apps. Each has its pros and cons. You should do your due diligence before jumping into custom app development to ensure that you are developing the best type for your needs and budget. To get you started in understanding the two types of apps, I want to talk about a few of the major differences between them.

Let's start with native apps since they are what most think of first. The major benefit of developing a native app is the ability to use the mobile features such as the sensors or the camera. However, you face a drawback if you lack a standard mobile device for all your employees. If that's the

KEY TERMS

Native app A mobile app developed to run on a specific platform and use that platform's operating system. Users download apps (free or for a fee) to their mobile device from a website. For example, iPhone apps are downloaded to the phone from Apple's App Store, and Android apps come from Google's Play Store (previously known as the Android Market). Native apps are developed specifically for each device.

Web app A mobile app designed to run on any mobile device. It is accessed using the mobile browser and the Internet. Web apps are most commonly developed in HTML.

case, then you must develop an app for each device. This would require either a development team or vendor familiar each device's development tools and standards. Also remember that native apps are downloaded to your mobile device through the app stores. Any apps that are acquired through an app store require you to review and approve them prior to launch. This additional step can be time consuming and may alter how you design the program to fit within your organization's apps development standards.

Companies often want to develop content once and deploy it across all devices. If this is your goal, then a web app may be the right answer. Web apps are not distributed through an app store; they are available on a website via your mobile browser and are typically developed using common web development tools. However, when designing the web app, remember that it normally can't access the unique device features.

The "develop once and deploy across multiple devices" concept is a huge benefit to web apps and is often what managers are looking for in a mobile app development project. A word of caution: make sure that your team thoroughly tests the program using a variety of mobile web browsers prior to launching it. This step is too often overlooked. Since each web browser is different, you may find some quirks based on the mobile web browser that is accessing the content. Don't assume your employees use the default program that comes with the device.

Using Mobile Learning to Gather Learner Data

Another way to use mobile learning is to gather intelligence and data about employees. You can do this in a variety of ways, ranging from assessing the knowledge they have retained from formal training to asking for participation in surveys to polling students. We discuss these three approaches next.

Assessments

Assessing if your employees have grasped the learning content and can apply it on the job should be a priority for any manager working on a training project, regardless of the delivery method. Before we jump into the mobile component, let's look at how we commonly assess our learners' knowledge obtained from traditional training methods.

During a classroom session, the instructor performs the assessment. This can take place by asking the students probing questions, having the students complete exercises, or giving them a formal test. With an e-learning course, it is common to give a formal test at the end of each lesson. In some cases, the course may be designed to require a cumulative test prior to giving the employee credit for taking the course.

Sounds like a solid approach, right? Well, by assessing the level of knowledge immediately following the instruction, we are really only relying on the employees' short-term recall of the information. However, the real question is: Do they remember the information and can they apply it a week later, a month later, or even a year after the course? Mobile devices can provide a means for assessing not only their short-term recall of the information but also long-term recall and even application of the content. The following are some basic techniques you can use to assess your employees' knowledge via mobile devices.

Pretests. A pretest is a short test given prior to the formalized event. This approach lets you gauge their existing knowledge on the subject matter and can assist you in identifying common knowledge gaps across your workforce.

Posttests. A posttest is a test given following a learning event to identify the extent to which students learned the content. By comparing the pretest results to the posttest results, you can identify if their knowledge on the particular subject has increased. Posttests can also provide you with data on where additional training content or approaches such as coaching or job aids can fill in the knowledge gaps.

Knowledge Checks. As a way to ensure that your employees retain the content well after the event, you should periodically send knowledge checks. One idea is to send them individual questions on their mobile devices. The beauty of this approach is that if one employee is struggling with a content area, you can provide additional content via the device to fill that employee's knowledge gap. Another idea is to send your employees a scenario and ask them to document how they would approach the situation or problem. Using speech recognition software on a mobile device is a quick way for the employee to return a response. By having the employee describe the solution in his or her own words, you can quickly identify what employees have and have not learned from the training.

Observation. Observing your employees while they are working is a good way to assess their ability to apply the training content. The employee can make his or her own observation and report to his or her manager, or the observation can be done by another employee. Here are a few ideas on how you can use mobile technologies to accomplish this task. While an employee is performing a task, have him or her take pictures at various stages of the work and submit the photos for review. Another idea is to have the employee use the mobile device's video camera to document a complete task from beginning to end. In both cases, you can review the artifacts (the photos or video) to ensure that the knowledge is applied on the job. Another idea is to use electronic checklists that a peer or manager fills in during a formal review. These checklists are easily completed using the device's touchscreen.

Surveys

Right or wrong, we often make decisions based on our perception of what employees need or want. Instead of guessing, use a mobile device to survey your employees on their opinions and needs. You can use a survey to ascertain employees' preferred learning method, content to cover in a course, new courses they need, opinions on whether a course helped them on the job, and even recommendations on how to improve a course.

Polling Students

Another means of gathering learner data (and to increase interactions with employees) is to poll them. A poll typically comprises multiple-choice questions in which you ask the group for input and they respond using an audience response system. When the poll closes, the software tabulates the responses and displays them on a PowerPoint slide for group discussion.

Until recently, you needed to give each participant an audience response device at the beginning of the session and hope they returned them when the session was over. You also were limited to multiple-choice or true/false questions. Today, software companies such as Poll Everywhere allow your employees to use their cell phones to answer the survey using text messages, Twitter, or even a web browser. This new technology also accommodates other types of polls such as freeform text where the participants can elaborate on their responses or ask questions of the instructor or manager.

Using Mobile Learning to Communicate and Collaborate

The techology of social media has great potential in the art of mobile learning. Let's explore how this technology might be used.

Social Media

Social media is a hot topic in training and provides unlimited opportunities not only for connecting people but for allowing them to democratize or self-regulate content. Social media comes in many forms, including blogs, wikis, microblogging, photo and video sites, Internet forums, social bookmarking sites, and social networking sites, to name a few.

KEY TERM

Social media Internet applications that enable communication and interactions among groups of people. Social media sites create a community where people share information such as personal profiles, opinions, insights, ideas, experiences, photos, and video. A few commonly recognized social media outlets are Facebook, Twitter, Yammer, Wikipedia, YouTube, Flickr, Pinterest, LinkedIn, and Delicious.

Learner-Generated Materials

When organizations usually think about creating training materials, they assume a top-down approach. The organization dictates the content

BANNING SOCIAL MEDIA

Social media raises many questions for companies. Is social media a distraction? What security issues arise if we allow employees to discuss business with outsiders? Do we want to be that transparent to the world? What do we do if employees speak negatively about the company? Often a company's first response (or as I think of it, a knee-jerk reaction) is to ban the use of social media in the workplace. Here are a few reasons why you may not want to ban social media.

- It's not the first time and will not be the last that new technology has been termed a workplace distraction. I've had many people tell me that when e-mail was new, many companies feared it would distract their workforce, yet today it's the lifeline of business communications.
- While your employees may be restricted from the social media sites while using the office network, there is nothing to stop them from accessing those sites from their mobile phones.
- Social media is a key component to how Millennials communicate. The fastest way to deter young talent from your company is by disallowing social media.
- You are losing a great opportunity to create an environment that encourages not only communication and collaboration, but also innovation.

Instead of banning social media, clearly communicate your expectations of appropriate use and find ways to embrace it for the good of the company.

and how it is created. Unfortunately, companies that subscribe to this approach miss the benefits of peer-to-peer learning. Most often the people who know the best way to apply content within the contexts of their jobs are the employees themselves. They have learned through experience what works best and how to troubleshoot when things go wrong. Instead of the top-down-only approach, I recommend that companies create an environment where they encourage their employees to share their expertise and knowledge with others. Social media outlets are a great way to do this.

Getting Innovative with Mobile Learning

There are a variety of ways you can create innovative mobile learning using social media technology.

Augmented Reality

I'm going to give you a few situations, and you decide if they are fact or fiction.

Imagine you're out of the country and unsure of that country's current currency exchange rate compared to U.S. dollars. You scan a $1 bill to find the exchange rate. Fact or fiction? Okay, next one. How about touring a history museum and the *Tyrannosaurus rex* display comes to 3D life right in front of you! Not as bones, but as how we believe it really looked. Fact or Fiction? Here's one more. You're in the market for a new house and while visiting a friend's house decided it was a great neighborhood to live in. You pick up your smartphone, circle a neighborhood on a map, and all of the houses for sale pop up on the map, complete with their selling prices. Fact or fiction?

> **KEY TERM**
>
> **Augmented reality** The enhancement of real-world experiences by overlaying digital elements with the physical environment. While looking at the physical environment through a mobile device with a camera (most commonly with your smartphone or special goggles), you will see overlays of data, images, video, music, and interactivity sitting on top of the real world.

Surprisingly, these are all real-life examples of augmented reality, blending our real lives with a digital world.

Augmented reality works in several ways. The first way is by creating a tag, which helps you categorize and filter e-learning content. When you point your smartphone or tablet at the tag, it comes to life, as with the example of the *Tyrannosaurus rex* coming to life in 3D. In this example, you download an app at the museum, and as you proceed with your tour, you will find tags posted on the wall. When you point your smartphone at the tag, the exhibit comes to life. The second way augmented reality works is by using image recognition software. When you point your device at an image or object, an app on your smartphone automatically recognizes what it is and presents the appropriate content. In the example of scanning your money, a Wikitude app includes image recognition software. This app provides hotspots on the image of the currency that allow you to find out more about the security features. The third way is through the use of location-based content using the camera on your device as well as sensors such as the GPS and the accelerometer on your phone. These apps determine exactly where you are and display the appropriate overlay data. The example of displaying the houses for sale in an area is an example of this technology.

As far as displaying augmented reality experiences, there are three options. The first display option is a head-mounted display such as a headset, goggles, or glasses. Contact lenses are currently in development and may be an option in the near future. The second display option is a handheld device such as your smartphone or tablet. The third option is a spatial display, where an image is projected onto a fixed surface.

Sounds cool, but can you use augmented reality as a learning tool? Absolutely. Remember that earlier in the book I shared a story about how I frequently use an app to learn about the stars? By holding my iPad toward the sky, I can see all the constellations, planets, and even the Hubble satellite. This app is augmented reality in action. Here are a couple ideas for using augmented reality with your mobile training.

First, you could add tags to your instructor-led materials to make them more interactive. Another idea is to turn your corporate campus into an augmented reality experience and add it to your new hire training program. A third idea is to use augmented reality as a performance support tool, providing your employees with step-by-step instructions and imagery while they are on the job. Imagine looking at a product with your mobile phone and having the troubleshooting steps appear on the screen.

Games

Did you know that 235 million people played games on Facebook between January 2012 and August 2012? Chances are, if you don't play, you know someone who does. While gaming has become a hot topic in training, there are a few myths that cause some people to resist using gaming in the workplace. When having a conversation with a manager about using mobile games, I often hear the following statements.

- Games are just for kids.
- Women don't play games, and about half my workforce is women.
- Games have no educational value and are strictly for fun and entertainment.

BMW Service Technicians Repair Cars with Augmented Reality

FOR EXAMPLE

According to BMW.com, BMW service technicians use augmented reality to obtain step-by-step instructions when servicing cars. The technicians wear a special pair of goggles that show them graphics and instructions while they are looking at the engine. For example: the program shows a technician how to perform each step of a task, and when the task is completed, the tech says, "next step" to continue the process.

The photo below breaks this down for you. In the upper right corner, you see the technician is tightening two screws. A virtual screwdriver shows the location of the first screw to tighten and, under that, you can see the technician's hands while he tightens the screw. A video on the BMW blog shows the technician completing a repair using augmented reality. The video can be found at www.bmwblog.com/2009/09/03/bmw-augmented-reality.

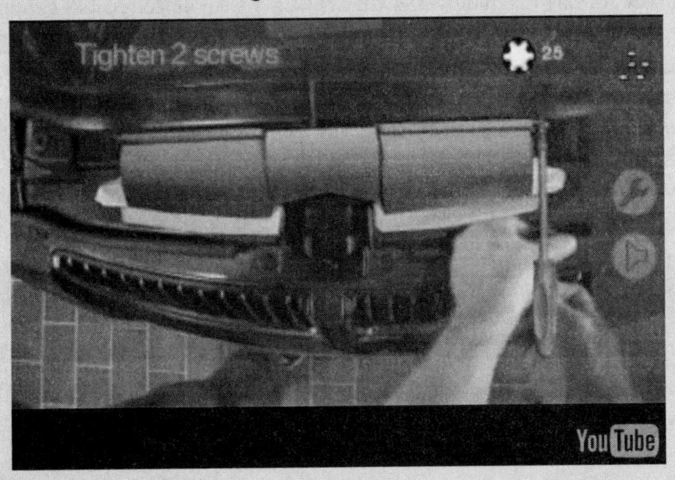

Service technician repairs a BMW with augmented reality.

Let's look at the age and gender myths first and get them out of the way. According to findings from the Entertainment Software Association (ESA), a game industry trade group, the average age of a gamer is 36 years old, and 29 percent of gamers are over age 50. And what about the myth that video games are played primarily by men? Surely that's a fact, right? According to ESA's findings, there is almost a 50–50 split between men and women. ESA also found that 42 percent of all U.S. women play video games.

Now that we've dispelled the age and gender myths, let's tackle the games-are-just-for-fun myth. Of course, games are fun. If they weren't, nobody would play them. And yes, they offer huge benefits in training our workforce. A few ways we can use mobile games include memorizing and performing tasks, assessing risks, problem solving, improving decision-making skills, and developing leadership skills.

FOR EXAMPLE

Learning Core Job Tasks with a Mobile Game

In each of our jobs, we perform some routine tasks that require us to memorize information or the steps of a process. Mobile games are an effective way to aid our employees in the memorization and application of the information. McDonald's in Japan is doing just that by teaching their new employees job tasks such as assembling hamburgers, frying french fries, and cleaning their workstations. They use a mobile learning game that runs on the Nintendo DS gaming system (see the photograph below), according to Mike Firm of Bloomberg News. In his interview with Hideki Narematsu, the HR manager of McDonald's Japan, Narematsu says, "People learn twice as quickly as compared to the old training methods, and they can apply those skills right away when they get into the workplace. When trainees come into the workplace they already remember the basic tasks and can spend more time building up their confidence in communication skills."

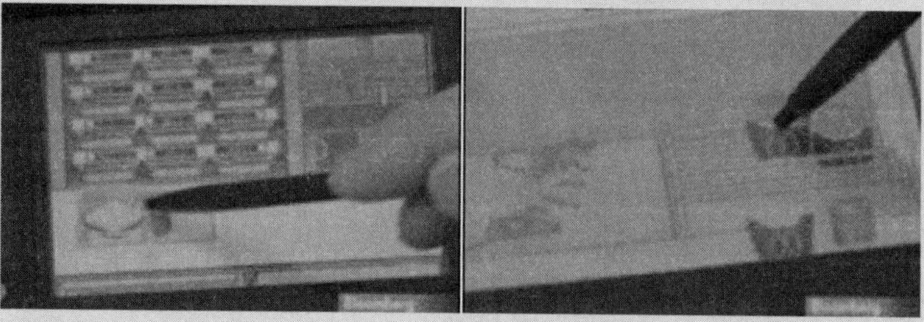

Mobile game teaches key skills to McDonald's Japan employees.

For those of you who want to read the research on the effectiveness of games in learning, there are many studies. Let's look at recent research performed by Traci Sitzmann, an assistant professor of management at the University of Colorado's business school, who spent a year researching the effectiveness of simulation games for adult learners by examining 65 studies and data from 6,476 adult trainees. Her research was published in *Personnel Psychology*, "A Meta-Analytic Examination of the Instructional Effectiveness of Computer-Based Simulation Games."

Sitamann found that when the games were designed as an active learning experience, included as a part of a larger curriculum, and offered the learners solid postgame debrief opportunities, they were effective teachers. When designed within these parameters, she found that students who learned through the games versus those trained using traditional training methods demonstrated an increase of 14 percent in skill-based knowledge, an 11 percent increase in their factual knowledge level, and a 9 percent higher retention rate. These are impressive findings.

DESIGN IS THE KEY TO GAME EFFECTIVENESS IN LEARNING

CAUTION

If you want to create mobile learning games, I strongly recommend that you include a game designer in the creation process and not attempt to do it on your own. Here are a few key components that make games effective. First, mobile learning games should comprise one component in your overarching training strategy. Don't create a game as the only training method, but blend it with other training opportunities. When creating games, it's easy to be distracted by the creative process and lose sight of the learning objectives. Second, make sure you clearly identify your learning objectives and how the game-play interactions support them. Even though it's a game, you still need to include the elements of any solid training initiative such as feedback and evaluation. Third, encourage your learners to play the game multiple times by motivating them intrinsically rather than through reward systems that motivate extrinsically and have only short-term benefits.

Immersive Learning Simulations

Think back to when you learned to drive. For most of us, we read the driver's manual and studied up on the rules of the road. When we felt confident, we took the written test to earn a driver's permit. It all seemed so black and white, so easy. But do you remember the first time you had to apply what you learned from the driver's manual to a real-world situation? Do you remember the first time someone cut in front of you without using his or her turn signal, or when a child ran in front of your car chasing a ball, or the first time you had a flat tire? Once you got behind the wheel of the vehicle, everything changed; it was way more complicated than the driver's manual made it seem. Today, many driver's education programs have incorporated immersive learning

KEY TERM

Immersive learning simulation The blending of simulation to re-create real-world experiences, pedagogy to ensure learning takes place, and game elements to fully immerse the learner into the goals of a learning experience.

simulations into their programs that allow the students to experience the complexity of driving before they actually hit the streets.

Historically, immersive learning simulations required large and expensive machinery to re-create the real-world environment. They were most often used to train employees who were working in hazardous or potentially life-threatening situations such as aviation, police work, healthcare, and military operations. Today, thanks to modern mobile technologies, these training solutions are now an option for a variety of training situations.

FOR EXAMPLE

Soldiers Train as a Platoon Using Immersive Learning Simulations

Today's soldiers are being trained using an immersive learning simulation called the Dismounted Soldier Training System. The program allows platoons of soldiers to train together in a virtual environment and run through real-life war scenarios. The training system is completely mobile. The soldiers wear special helmets to see and hear within the virtual environment. Their uniforms and weapons are equipped with sensors that communicate the soldier's movements and actions to their avatar operating in a virtual environment. You can view a video of this training on YouTube at www.youtube.com/watch?v=AMyoQhUcPgM.

Soldier training with Dismounted Soldier Training System

Some of the benefits of immersive learning simulation are:

- Increasing learner motivation
- Learning from your mistakes in a safe environment
- Practicing scenarios that would be impossible to train for in the real world
- Integrating knowledge and skills
- Assessing employees' knowledge based on their actual performance with the simulation

Manager's Checklist for Chapter 4

- ☑ When beginning to think about leveraging mobile learning in your organization, start by focusing on the user experience instead of the technology.
- ☑ People use their mobile devices in short bursts. Mobile training and performance support should be designed in the same fashion.
- ☑ Mobile learning is a viable way to provide information to your employees in a variety of formats including videos, podcasts, e-books, job aids, reminders, alerts, and mobile applications.
- ☑ Mobile learning can provide valuable data through surveys, assessments, knowledge checks, and polling.
- ☑ Encouraging the use of social media from the beginning will result in increased communication and collaboration.
- ☑ Social media is a great way to encourage your learners to generate content materials and share their knowledge and best practices with other employees.
- ☑ Use innovative approaches to mobile learning such as gaming, immersive simulations, and augmented reality to create a highly engaging and interactive experience that teaches and reinforces critical business skills.

The Mobile Learning Vision Statement

Now that you have an understanding of the mobile learning landscape and its viable uses, we discuss how to document your vision for mobile learning. This chapter focuses on your first step in creating a business case: to document your mobile learning vision statement. Your vision statement should include documenting your current state, articulating your desired future state, explaining your mobile learning rationale, and identifying how you'll define as well as measure the success of your program.

Documenting Your Current State

The first step in the process of creating your business case and your mobile learning vision is to understand and clearly articulate where your organization is today. This includes explaining what the problem is that you'll be addressing, what is causing the problem, and who are the key players in fixing the problem.

Identifying the Problem

The first step in defining your mobile learning vision is to articulate the problem you're trying to solve. Start with the high-level problem or issue that your project will address. Some of the most common problems mobile learning can help solve include:

The Five Ws and One H Technique

A technique I use with my clients when I help them in craft their mobile learning vision is to apply the "Five Ws and One H" approach. Journalists use this approach when they're preparing for interviews and writing their articles. I used this approach when I was a writer for the *St. Louis Post-Dispatch* and *St. Louis* magazine. When I switched careers to training and development I found that this approach worked well for managers and executives when writing a business case as well as for the development teams when they wrote the training content. In a nutshell, the Five Ws and One H are simply a series of questions that you ask and answer. These questions are What, When, Why, Where, Who, and How. As you're documenting your business case, ask yourself appropriate what, when, why, where, who, and how questions. Here are a few examples of the types of questions you should be answering for your business case.

- Why am I pursuing a mobile learning solution?
- What is the problem I am solving?
- What will the solution be?
- Who is my audience for the solution?
- When do they need learning content?
- Where are the employees when they need performance support or additional information to improve their job performance?
- How will I roll out the program?
- How will I measure its success?
- What will the program cost?

By using the Five Ws and One H approach, you'll ensure that you've developed a well-thought-out plan that will assist in gaining stakeholder approval as well as assist the Learning and Development team in meeting the programs goals.

- Increasing sales
- Increasing employee productivity
- Improving employee accuracy on the job
- Decreasing the number of mistakes
- Reducing the number of safety incidents
- Increasing communication and knowledge sharing among employees
- Increasing the effectiveness of your training through spaced learning

Let's break that down into a few tangible examples. Are your service technicians taking too long to troubleshoot and fix common problems? Is your sales force struggling to address common obstacles with their sales process and losing sales? Are all best practices being shared throughout the company so all employees can benefit from your top performers' wisdom? When deploying new equipment in the field, do you experience spikes in the number of employee accidents? Is all of your training on a topic crammed into one formal course, resulting in employees forgetting what they learned in training before they get back to their jobs?

Know What You're Solving

CAUTION

When I speak to managers and executives about why they're looking to get into mobile learning, I often hear that they're looking for a new and innovative approach to employee training. Two of the most common problems managers face when implementing a mobile learning solution are (1) not having a solid grasp of the problem they want to solve and (2) not having a solid business case in place prior to launching a pilot project.

All too often managers will jump into a pilot project to test the waters for the acceptance of mobile learning without having a solid rationale for why they're doing so or even how they'll know if the project has been a success. To set your project up for both corporate-wide acceptance and for a successful business outcome, take the time to document your business case and ensure that mobile learning will in fact solve the problems your organization is facing.

The next step is to identify what is causing the problem. Is it a breakdown in your business process? Are people not receiving the right training at the right time? Is your training content outdated? Do your employees lack the right tools to achieve peak performance when their demanding jobs expect them to do more with less? Identifying all the root causes of your problem is critical to ensure that training can help you achieve your goals.

Once you have identified the high-level problem and all of the causes, then you can begin to quantify the problem. Here are some examples of high-level problems that mobile learning can address.

- The service technicians are taking an average of two hours per service call to troubleshoot and fix problems that should only take one and a quarter hours.
- Our help desk consistently receives the same question on an average of 20 times per week.
- On 15 percent of our service calls, we need to return at least three times to identify and fix the problem.

Can Mobile Learning Solve the Problem?

Companies and even the Learning and Development teams will often want to immediately update training when a problem has been identified. Before you jump into crafting your mobile learning solution, make sure that you have thoroughly analyzed the root causes of the problem. While training may be a component, often the root causes are at a deeper level, such as internal systems that needs to be streamlined or business processes need to be updated. Once you have identified how you will fix those underlying key issues, then you can think about how to leverage mobile learning.

- We are losing approximately $200,000 in new sales each quarter by being unable to address our prospects' objections.
- Over the last two years we have experienced a 25 percent increase in the number of safety incidents reported per site when we deploy new machinery compared to existing machinery.

For simplicity, the examples here are single-impact statements, but when you are documenting your problem, make sure you look at the potential effects on all business units as well as the total organization.

SMART MANAGING

Honesty Counts

When defining your current state, you should be brutally honest regarding the current issue or problems your company is facing. Often managers don't want to share the depth of the problem with others inside the organization for fear of retribution. However, if you don't document *all* the issues and challenges, you will be unable to create a mobile learning solution that will solve them. The executives and your Learning and Development team must be aware of the depth of the problem to ensure adequate funding and an effective mobile learning solution.

Identifying Your Audience

Now that you've got a good handle on the problem you are solving, your next step is to identify who your primary audience is and their readiness for a mobile learning solution.

Identify all the employees who could possibly contribute to solving the problem. I recommend that you categorize your audience into two groups: the primary audience and the secondary audience.

Primary audience The target group of individuals for your learning solution. This group will receive the primary benefit from the mobile learning experience.

Secondary audience Everyone who interacts with your primary audience and contributes to your primary audience's success in achieving the business goals. This could include managers, administrative staff, help desk employees, and even your Learning and Development team.

QUESTIONS TO ADDRESS REGARDING YOUR AUDIENCE

When documenting your high-level audience analysis, answer the following questions.

- Who needs training to solve the problem?
- How many people are in the audience?
- Why do they need the training or performance support?
- What do they need to know in the future to address the problem?
- How do they receive their training today?
- When do they need the training or access to performance support?
- Where are the employees when they need the information (in the field vs. at their desks)?

For example, if you're working on a mobile solution that will impact your company's sales numbers, the primary audience might be your sales representatives and even possibly your customers if you intend to create learning content specifically for them. Your secondary audiences might include the sales division's senior management, regional sales managers, department managers, and the sales support team.

Let's start with understanding your primary audience. Remember when defining your *primary audience* that these are the individuals who will benefit most from the mobile solution and for whom you are designing the solution. As you identify each group of employees who fall within your primary audience, you should document the audience size, geographic locations, and basic demographics, such as sex, age ranges, and average time with the company. Your HR department should be able to pull reports to help you gather this information.

Once you have a grasp of who could be involved in solving the problem, you should now gain an understanding of their experience with and readiness for accepting mobile technologies. Your goal here is to create a

TRICKS OF THE TRADE

Generational Analysis

When you're looking at how your employees are using or will accept mobile technologies, look at the generational groupings (Millennials, Gen X, Baby Boomers, and Traditionalists) for a better insight into the trends and needs of your employees.

Typically when companies create age groupings, they classify people within six age ranges: 21 and under, 22–34, 35–44, 45–54, 55–64, and 65 and over. The problem with this approach is that multiple generational groups fall within multiple ranges. For example, in 2013 the age range for the Gen X group spanned 33–48 years old. If we look at the standard ranges, that group includes a portion of the 22–34 set, all of the 35–44 set, and a portion of the 45–54 set. With the standard approach there is no way to clearly identify and separate the preferences of your Gen X employee population.

To allow you the flexibility of analyzing your audience from a generational perspective, you should use the year an employee was born, not his or her current age. The born-between ranges would be: After 1980 (Millennial), 1965–1980 (Gen X), 1946–1964 (Baby Boomer), and 1928–1945 (Traditionalist).

primary audience profile to ensure that your mobile solution will resonate with your audience and solve your problems.

One of the easiest ways to gather this information is by surveying your employees. Software such as Survey Monkey helps you quickly set up, distribute, and automate the results. Don't feel as if you need to interview or survey all your audience members (which could feel like a daunting task), but rather, identify a representative group. Your representative group should consist of employees similar to your audience profile. The larger your audience base, the greater the number of people you should survey. Here is a sample readiness survey you can use.

Sample Mobile Readiness Survey

While the specific questions you'll ask can vary, the following is a potential list of survey questions that you can use to get started in gaining information on the mobile preferences and readiness of your audience.

Mobile Readiness Survey

Job Title ______________________________

Which Age Range ________________________

Gender ________________________________

What type of mobile phone do you currently use?

- ❑ Cell phone (non-smartphone)
- ❑ Smartphone (please select the type)
  - ❑ Android
  - ❑ BlackBerry
  - ❑ iPhone
  - ❑ Windows
  - ❑ Other (fill in the blank) ______________________________

I use my mobile phone to (select all that apply):

- ❑ Communicate with others via:
- ❑ E-mail
- ❑ Text messages
- ❑ Phone calls
- ❑ Listen to podcasts
- ❑ Get navigational assistance (directions/maps)
- ❑ Play games
- ❑ Search the web for information
- ❑ Use apps (list your top 5 apps)

  ____

  ____

  ____

  ____

  ____
- ❑ Watch videos

What type of tablet device do you use?

- ❑ Apple iPad
- ❑ Galaxy
- ❑ Playbook
- ❑ Windows Surface
- ❑ Other (please list) ______________________________
- ❑ I don't own a tablet

List the top 5 tasks you would prefer to use a tablet for instead of a laptop while you're working away from your desk.

____

____

____

____

____

If they were available, I would use my mobile phone to access company information, training materials, and performance support tools.
- ❑ Yes
- ❑ No
- ❑ Unsure (I would have to experience it.)

If they were available, I would use my tablet to access company information, training materials, and performance support tools.
- ❑ Yes
- ❑ No
- ❑ Unsure (I would have to experience it.)

Now that you have a good understanding of your primary audience, you should identify your secondary audience (employees who support or contribute to your primary audience's success in achieving their goals). When documenting your secondary audience, identify how and when that group contributes to your primary audience's success. You only need to perform a mobile readiness survey if this audience will interact directly with the mobile learning solution.

Crafting Your Desired Future State

Now that you have clearly defined the problem that you want to address and you understand your primary and secondary audience, you can define your desired future state. Your future state answers the following two questions: Where do you see the business in the future? How will you get there?

Let's look at a couple of examples. Let's say you're addressing the challenge of reducing the time service technicians spend to troubleshoot and fix common problems from two hours to one and a quarter hours. Your future state definition could be the following:

> Service technicians will take an average of one and a quarter hours to repair common problems in the field. We'll reach that goal by identifying the most common problems and providing our team with performance support tools that they can use in the field to quickly troubleshoot problems. Once a problem is identified, the mobile performance support tools will walk our technicians through each step of the process as they complete the task. The employees will use their personal smartphones, and the program will be developed for iPhone and Android devices.

Notice how in this example we did not scope out the complete project. We defined a high-level vision of the new, desired, future performance state and how we will address the problem.

Let's look at another example. Let's say that you're addressing the problem of helping your customers make an educated decision at the point of sale, with a goal of increasing sales. Your desired future state description could be as follows:

> We will increase consumer awareness of which product best fulfills their needs as they're making the buying decision in the store. We will accomplish this goal by adding QR codes to our store signage. The signage should also include a statement on where customers can download a QR scanner if they do not already have one installed on their smartphones.
>
> When customers scan the QR code, a web page will launch. This web page will walk our customers through a series of questions and, based on their responses, will recommend which of our products best meets their needs. After completing the questionnaire, we will also provide them with a link to a 20 percent off coupon for their first purchase of the product. The customer will only need to hand the smartphone to the sales clerk who will scan the code at checkout.

Notice how we did not determine at this point the total number of questions that we needed to ask or the details of how we would track the first purchase. Both of those points will be determined later. The important point is that you have a vision that you can clearly articulate to the IT stakeholders and the Learning and Development team who will figure out all the details of how to accomplish your vision.

Don't Get Caught Up in the Details

CAUTION

Often managers feel the need to flesh out all the details in their mobile learning vision. They believe they need to know all the content to be covered and how the program will be specifically designed and tracked. They may start wondering about how IT will handle their requests. When managers craft a vision statement, they often want to explain how the system will look. Fight off those urges and focus on explaining what the future state will look like and how mobile learning or mobile performance support fits into the picture. Make sure your vision has enough details so others understand it, but don't feel obligated to fill in all the blanks. That's why you have a team of stakeholders and other departments to help you.

Describing Your Mobile Learning Rationale

Up to this point, you've identified and documented the problem you're solving. You've thought through who your primary and secondary audiences are, as well as their readiness for a mobile learning solution. And you have crafted a vision of what the future will look like and how mobile learning fits into that picture. Now it's time to think through and document why a mobile learning solution is the best fit for your organization. In this section, you'll want to address your audiences' readiness for a mobile learning solution, identify when and how the solution fits the needs of your employees, and the benefits of your mobile learning solution.

QUESTIONS TO ANSWER WHEN DESCRIBING YOUR RATIONALE

TOOLS

- Why is mobile learning the best delivery method?
- How does mobile learning support your workforce in meeting the organization's goals?
- How does this mobile learning project augment or support other training initiatives on the topic?
- What training content (if any) will mobile learning replace in your organization and why?
- When do you envision your employees or customers using the mobile learning solution?
- How will the company benefit from the mobile learning project?

Audience Readiness

Earlier, you surveyed your employees to find out what devices they have and how they use them in their daily lives. Now you'll share those results as they relate to your proposed solution. When possible, use graphs to show the results.

The following are a few ways in which you can look at the results of your mobile readiness survey. How many people in your target audience already have mobile devices and which ones? If most of your primary audience uses an iPhone or Android, then that could be a justification for developing a solution that only plays on those two devices. Are people already employing their devices for personal use in a fashion similar to your proposed mobile learning solutions? For example, are they already watching videos, using productivity apps, or playing games on the

What If My Findings Show the Employees Are Not Ready?

CAUTION

You may encounter an audience that is not ready for a mobile learning solution. Does that mean you should forget about your mobile learning vision? In some cases the answer may be yes, but most often the answer is no. Use this insight to determine what you need to do to prepare your employees for mobile learning and devise a plan to do so. If your employees do not currently have smartphones or tablets and the phones are a key factor in your solution, then maybe the company will need to invest in mobile devices. If the employees don't see how mobile learning can make their jobs easier, then you will have to educate them.

Remember, just as you probably envisioned mobile learning as taking your existing classes on a phone, your audience probably believes that as well. I suggest you ask yourself: How can we prepare the audience for a successful mobile learning experience? If your answer is "we can't," then mobile learning is probably not the right solution; however, in most situations, that is not the case.

devices? What percentage of your audience is open to using their personal devices for mobile learning if it became available? How many of your employees fall within the Millennial generation who expect to have mobile solutions available to them in the workplace?

How Does Your Mobile Learning Fit Your Employees' Needs?

Earlier in the book, we discussed Mosher and Gottfredson's model, the Five Moments of Need, which are the moments that motivate your employees to learn on the job. Do you remember what those moments are? Just in case you can't remember all five, here they are again:

1. Learning how to do something for the first time (new content)
2. Expanding on the information your employees already have learned (more content/context)
3. When they need to act upon what they have learned (application)
4. When problems arise (solve)
5. When employees need to learn a new way to perform in their jobs (change)

The first two moments of need lend themselves to formal training events, and the last three moments of need lend themselves to informal learning events.

Now that you have refreshed your memory on the moments of need, take some time to think through which of these moments your mobile learning solution will address and explain how mobile learning will fulfill the need(s). Remember: if you are solving a complex problem, you may need to cover multiple moments of need. For example, you may have to provide new learning content if there is a change in a business process and then provide performance support as your employees change how they perform their daily tasks. When this is the case, make sure that you identify each learning need and discuss it as a separate item.

Identifying the Benefits of Your Mobile Learning Solution

Now that you have really thought through how your mobile learning experience fits each of your employees' moments of need to solve the problem, you document the benefits that you expect to achieve by using this delivery method. For each mobile learning element, document the benefits of the delivery method versus other training options (i.e., instructor-led training, e-learning, distance learning, paper-based job aids, mentoring, etc.). While you are documenting the benefits, keep in mind some of the ones we've already discussed such as:

- Providing anywhere and anytime learning opportunities
- Providing content and information just in time (when the employee needs it the most)
- Providing location- and context-specific content
- Collecting data (both for the employee and for the company)
- Increasing the effectiveness of your formal training by extending the learning process

Defining and Measuring the Success of Your Mobile Learning Initiative

Now that you've thought through and documented why your mobile learning solution is the best approach, you are ready to move to the final step in your mobile learning vision statement. This final step is all about how you and others in the organization will be able to determine if the

THERE ARE MORE THAN FIVE BENEFITS TO MOBILE LEARNING

CAUTION

Throughout the book we have discussed the potential benefits of mobile learning. While the benefits discussed here focus on the unique benefits of mobile learning versus other training delivery methods, the list is not complete. Let's discuss a few examples of case-specific benefits.

1. If your company is deploying a new software program that runs from a mobile device, then providing training on the actual device is a benefit.
2. As a means of attracting younger talent, demonstrate how you can fulfill their needs for mobile content and provide a mobile-friendly working environment.
3. Use mobile learning as a differentiator from how your competitors handle the problem.

Remember to identify all the benefits that your organization can realize through the mobile learning solution and not only those we have touched on in this book.

project was a success. Executives commonly equate a successful project with a calculated return on their investment. That is not the focus of this chapter, although we do discuss that subject later in the book. For now, we are looking at a wider array of success factors that focus on the employees and the impact on the organization.

How to Measure Training Effectiveness

When it comes to measuring the effectiveness of training solutions, one of the most commonly used models is Donald Kirkpatrick's Four Levels of Evaluation, outlined in his book *Evaluating Training Programs,* first published in 1994. For most training and development professionals, his theory is the holy grail in determining if the training program was effective and a successful endeavor. Review Table 5-1 to gain a high-level understanding of Kirkpatrick's four levels of evaluation. Then we will discuss each level.

The four levels of evaluation give you a framework for evaluating a training program. The first assumption people make when they hear of the levels of evaluation is that a student must complete one level before moving on to the next. That is not the case. What you measure is project specific. For example: if you are deploying a mobile learning performance support tool, you do not need to measure if there was an increase in

Level	What Is Measured	Common Ways to Measure
1: Reaction	Student satisfaction	Surveys
2: Learning	Increase in the learners' knowledge	Pre- and Posttests
3: Behavior	Application of the learning content on the job	Observation of the employee on the job (over time)
4: Results	Impact on the organization	Reports from internal systems and/or calculating your return on investment

Table 5-1. Summary of Kirkpatrick's Four Levels of Evaluation

knowledge (Level 2). This may seem illogical to you, but as you read the following sections, it should become clear.

One note before we get into the details of the four levels of evaluation. The intent of this section isn't how to perform these evaluations, but to give you the information you need to define how you will measure a successful mobile learning solution. You need to work with others in your organization such as IT, your Learning and Development team, or vendor to flesh out the details.

Measuring Your Employees' Reactions

Let's look at Level 1, which measures the learners' reaction immediately following the completion of the training program. Think of this level as a way to measure how your employees feel about the learning event. You measure their reactions by asking them to complete a survey, which consists of a combination of questions that the student answers using a Likert scale and a few short-response questions. Typically, the survey consists of no more than 10 questions.

Level 1 evaluations are a staple of instructor-led training classes and e-learning courses. Some of the student reactions that a formal training event survey can address include:

1. Were the goals of the training clearly defined?
2. Were the topics covered relevant to the course?
3. Were the learning objectives clear?
4. Will the student use what he or she learned on the job?

5. How often will the student use the content on his or her daily job?
6. Was the content effectively communicated?
7. Did the student have sufficient practice opportunity?
8. Was the duration of the class appropriate?

KEY TERM

Likert scale Typically a five-point scale to measure participants' attitudes such as level of agreement with a statement, frequency of use, importance to them, or likeliness of use. The five points on a Likert scale measuring agreement would be Strongly Agree, Agree, Neutral, Disagree, and Strongly Disagree. If you do not want the ambiguity of neutral, you can modify this to a four-point by scale eliminating the Neutral response.

When surveying your employees on their reaction to mobile learning, you will need to work with your training team or vendor to devise questions that will give you worthwhile information. Remember

FOR EXAMPLE

Example of a Formal Learning Event Level 1 Survey

The following is an example of a standard Level I survey that you might take after attending a conference session. Notice how the questions are phrased as statements versus yes/no responses. Also note, at the bottom of the survey, there are two questions that allow the participant to voice feelings about or reactions to the session in his or her own words.

Questions	Strongly Agree	Agree	Neutral	Disagree	Strongly Disagree
1. I found the session a good use of my time.	5	4	3	2	1
2. The session met the learning objectives.	5	4	3	2	1
3. The facilitator was knowledgeable.	5	4	3	2	1
4. The facilitator involved the participants.	5	4	3	2	1
5. I will use the content learned in the session.	5	4	3	2	1
6. What did you like best about the session?					
7. What changes would you recommend?					

that your training team should develop questions that relate to the goals of your mobile learning project and vision.

DEVELOPING A MEANINGFUL MOBILE LEVEL 1 SURVEY

While you won't be responsible for writing your mobile learning solutions' Level I surveys, you should ensure that the surveys give you valuable data. Don't accept a question that gives meaningless information such as: Did you like the mobile learning experience? Insist on questions that relate directly to the problem you need to solve. For example, ask students if the mobile learning experience helped them overcome the customer's objections during the sales cycle or if it helped them realize the potential safety hazards of using a new piece of machinery. Remember to spin these questions into statements that employees will answer using a Likert scale versus soliciting simple yes/no responses.

Measuring If Learning Occurred

Another staple in a learning program is to measure if the students actually increased their knowledge, changed their attitudes, or increased their skills by participating in the training program. The concept is to provide a pretest to measure your employees' current knowledge and immediately follow the training with a posttest. The posttest lets you see if your employees' knowledge increased because of the training by comparing the results of the two tests.

While some companies test their audience's knowledge before and after the training event, often companies only test after the event. The questions in the posttest are based on the course's learning objectives and do not measure an increase in knowledge, but are a way to determine if the student can move on to the next content block or receive a course grade. The assumption is that if the student can move forward or pass the course, then learning has occurred. However, without understanding an employee's current knowledge level prior to the training event, there is no way to measure that knowledge increased due to the event. Perhaps the employee already had the knowledge and could have passed the test without taking the course.

The key to measuring if learning occurred in the training event is to measure three areas.

1. What tangible knowledge did the learner gain due to the training event?

We Do What We Know

SMART MANAGING

When defining how you'll measure your program's success, many managers who are new to mobile learning fall back to what they have been accustomed to in other training programs: offer a course and afterward test the student. Before you jump on the testing bandwagon, you should evaluate if a test is a valid measure. Ask yourself what kind of solution you envision for mobile learning. Are you providing performance support assistance to your employees? If so, a test is not the right approach. Are you providing content at the exact moment when the learner can use it? If so, a posttest may not be your answer to measure successful training. Remember, you don't need to evaluate all four levels to have a successful project.

2. What skills did your employees either develop or improve due to the training?
3. What attitudes in your employees were modified because of the training?

In all cases, the key is to measure increased knowledge due to the training event. Often people only think about measuring the tangible knowledge acquired and forget about the skills and attitude component to this level. To measure skills, give a performance-based test or simulation instead of a standardized test. For attitudes, third-party observation is one of the best methods.

Tests in Training: Do They Truly Measure Knowledge?

MISTAKE PROOFING

We may believe that if a student passes a test, she or he has learned something. However, think about your average test. It usually consists of true/false and multiple-choice questions. These types of tests do not really assess true knowledge; they assess a student's ability to recall the correct answer from a list of possible answers. In addition, if we test only immediately following the course, the student is using short-term memory, which fades over time. When testing, think out of the box on how you can truly assess if the student has increased her or his knowledge and skills and changed attitudes. Test the student based on performing a task or using the information in the context of the job. Performance-based assessments or simulations are more effective than standardized testing at assessing your employees' knowledge.

Changing Behavior

From a business perspective, we want our training programs to have a positive impact on the organization, and the first step is seeing a change in our employees' daily behaviors. The first thing to consider when measuring behavioral change is that this is not a measure-once-and-never-again proposition. I recommend that you evaluate shortly after the training event and then measure at several intervals. After all, behavior change is about creating a new habit, and the average adult needs about 45 days to truly make new behavior a part of her or his daily life without falling back into old routines. Measuring over time allows you to have confidence that the student has achieved sustainable change.

Evaluating and measuring your employees' behavioral change are often accomplished through a combination of face-to-face observations of the employees performing their jobs and personal interviews with the employees. By combining these two elements, you can see tangible change as well as the employees' perceptions of their changed behavior.

Questions to Ask When Measuring Behavior Changes

When measuring behavioral change, consider the following questions about your employees:

1. Are they using their new knowledge and skills on the job?
2. Are you seeing a measurable change in employee performance?
3. Can they explain their knowledge to other employees?
4. Do they continue to exhibit the new behavior over a period of time or have they fallen back into the old routine?

For observations, tools should be created for the employees' managers or team members to use while they are making their observations. These tools could include checklists to use while the employees are performing their jobs or scenarios that employees could be asked to perform. These tools will eliminate any chance of subjective evaluation and should align with the desired performance criteria or indicators. Immediately after the evaluation, provide feedback to the employees on their performance to raise their awareness and address any problems.

The Necessary Ingredients for Change in the Workplace

SMART MANAGING

Four key ingredients are needed for employee behavioral change.

1. Employees must be willing to change their behavior. If they lack the desire and motivation to change, then you have hit an impasse.
2. As a manager, you need to make sure employees know what to do, when to do it, and how to do it. You must also provide them with the tools needed to perform the tasks. This is where mobile learning solutions fit into the picture.
3. You must provide them with the right work environment and conditions.
4. Often forgotten, you must reward them when they change.

Rewards need not be major incentives or even financial. They can be as simple as complimenting your staff or recognizing them in front of their peers. Company social networks that can be accessed through the employees' mobile devices can be a great way to share your employees' achievements. Remember, before you jump into changing your employees' behaviors, make sure you have all the necessary ingredients.

Measuring Results

We all want our training programs to positively impact the business. For some solutions, you will not need to perform an in-depth analysis to identify the precise impact that the mobile learning program has had on the business. Mobile learning designed to aid performance and change behaviors should have a positive impact on the business. If you measure Level 3 and confirm change has occurred, then you should be seeing the results in action. We are not talking right now about calculating your return on investment; that is a different step, and one we cover in depth in Chapter 10.

The results we are measuring at this point should align with the problem you're solving, such as increasing sales, reducing safety incidents, increasing productivity, or improving your employees' work quality. This is where your training solution faces the acid test: to see if you achieved those goals you identified in your vision statement. The first step is to recall the goals you identified earlier in your desired future state and to identify all the performance indicators you need to measure. Examples of the performance indicators could be:

- Total number of sales
- Amount of time to close sales
- Number of safety incidents reported
- Severity of safety incidents reported
- Time to fix problems
- Waste figures
- Quality ratings
- Rework figures
- Number of customer complaints

Once you have identified the performance indicators, you need to identify where this data will come from. Do you have systems and reports in place today that capture the information, or do you need to work with your IT department to create custom reports?

TOOLS

Definition and Measurement Questions to Answer

The following are some questions you should consider when documenting how you will define and measure the success of your learning solution. If you're defining success on multiple levels, you should answer these questions for each way you'll determine if the mobile learning solution is successful.

- How will you the define success of your mobile learning project?
- What will you measure?
- Where is the information or data coming from to allow you to measure the program?
- When or at what points in the program will you measure success?
- Who will be responsible for obtaining and reporting the results?
- How will you communicate the success of the program?

When do you begin to measure the results? For a large-scale training initiative, I recommend you give your training program about three to six months before you measure the organizational impact. Remember that it takes time for your employees to change their behavior and that is what is required to see results. You should measure the program not once, but rather, a few times throughout the year. This will allow you to rule out outside influences (such as seasonal variations in your business or even chance) so you can attribute your employees' success to the training program and not to outside influencers. Note that when multiple influencers affect the employees' ability to achieve results, such as a change in a business process or deploying improved systems, you face a challenge in isolating how the mobile learning solution impacted the total results.

Bringing It Together

Let's talk about these levels in regard to mobile learning. Remember that the problem you're addressing, the type of mobile experience you're creating, and whether this is your audience's first mobile learning experience all affect what you measure. Regardless of what you measure, you must define your success criteria and how you will measure the effectiveness of the solution prior to developing your solution.

Mobile learning is more about augmenting your training programs, providing additional context to your employees, and empowering them with performance support than a formalized training program. Make sure that you define success by how you will leverage mobile learning in your organization and for the type of solution you will deploy.

TRICKS OF THE TRADE

WORK BACKWARD

When working on defining a successful project, I recommend that you start at Level 4 by identifying the results you want to achieve and how you will measure success. From there, you can think through and define how well the modified behaviors achieved the desired results. Once you have identified the behaviors, a natural jump is to think through what knowledge, skills, and attitudes need to be enhanced. From there, you move into measuring your employees' feelings and reactions to the mobile learning solution. Remember, you do not need to define success in each area; you must make sure that whatever you use to define your project's success can be measured and can demonstrate a successful mobile learning project.

Manager's Checklist for Chapter 5

- ☑ Before moving into a mobile learning project, you need a solid vision that you can communicate to executives, IT, and your Learning and Development team or mobile learning vendor.
- ☑ Ensure you have a handle on the business problem that your solution will address, including how to quantify the impact the problem has on all affected business units or the organization.

- ☑ Understand who composes your primary and secondary audiences as well as how their knowledge, skills, and attitudes impact the problem. Understand how they currently use mobile devices in their personal lives and their readiness for mobile learning.
- ☑ Clearly document your vision for how using mobile learning can address your current business problem and what the experience will be like for your audiences.
- ☑ Take time to ensure that mobile learning is the best way to solve your problem. Document your rationale for why this is the case.
- ☑ Define your success criteria and how you will measure the success and effectiveness of your mobile learning solution. Areas to analyze include the learners' reaction to the mobile learning; their increase in knowledge, skills, and attitudes; their change in on-the-job behaviors; and the quantifiable results from a business perspective.

Requirements and Expectations for Mobile Learning

Once you have documented your mobile learning vision, including identifying your current state and future state, outlining a mobile learning rationale, and defining a successful project, you can move into determining your program requirements and expectations. In this chapter, we focus on two distinct sets of requirements you'll need to account for in your plan. First, you must identify the stakeholder requirements, and second, you must identify the content requirements for your mobile learning solutions. During this requirements gathering step, depending on the size of your organization, you may have team members who will perform this work. However, you should understand the key components of the process. Next, we discuss the environmental and situational considerations when deploying a mobile learning solution. Last, we discuss anticipating and planning for the changes in your organization.

Identifying Stakeholders' Requirements

The first step you'll take to define your mobile learning project requirements is to identify your project stakeholders. When listing all your mobile learning project stakeholders, ask yourself: Do these individuals have the control and the authority to approve or veto components of my mobile learning project that are specific to the group they represent? If they do, then make sure you include them in your stakeholder list. Since these peo-

KEY TERM

Stakeholder An individual or a representative from a larger group who has a direct influence on your project or is directly affected by the outcome. As such, he or she has a "stake" in the success of the mobile learning initiative.

ple have a direct "stake" in the success of the mobile learning initiative, it's critical that you have a solid understanding of their requirements. Remember to include yourself on the stakeholder list.

MISTAKE PROOFING

PULLING TOGETHER THE ALL-STAR TEAM

A frequent mistake that managers make when identifying their stakeholder requirements is that they do not include all the stakeholders up front. There is nothing worse than having developed your mobile content and being ready to go live only to find, for example, that the legal department needed to review it. Here are a few questions to ask when identifying your stakeholders.

- Have you included a representative from the IT department?
- Do you have a senior executive as a champion?
- Have you included an HR representative?
- Do you have representation of senior managers from all the affected business units?
- Does the legal department need to be involved?
- Do you have representatives from your primary audience?

The key is to remember that your stakeholders include anyone who can veto your solution.

Now that you have identified all the project stakeholders, it's time to sit down with them and scope out all their requirements. When discussing their requirements, be sure you ask them questions regarding their specific goals for success, their views on the mobile learning initiative, risks they see in implementing the solution, and what short- and long-term limitations they foresee with using mobile devices specific to the group they represent. When meeting with the stakeholders, discuss the level of communication they'll need. This conversation is also a great opportunity to share with them the role that you foresee them playing in the project's development. Their project responsibilities may include:

- Owning the content
- Providing feedback and approving critical decisions

- Providing the budget
- Providing the resources to be used on the project
- Providing a clear path for escalating issues and resolving problems
- Increasing program awareness within the group they represent
- Evangelizing the initiative within the organization

TRICKS OF THE TRADE

ENSURING ALIGNMENT ON REQUIREMENTS

One of the best ways to gather your stakeholders' requirements is to have an active conversation with them. However, sometimes stakeholders may go off on a tangent and a key point will be lost. To ensure that you fully grasp their requirements and that your stakeholders don't omit anything, I recommend that you resubmit the requirements to them in written format. Ask them to review the document to assure that you have addressed all their concerns. When you reach agreement on their requirements, ask them to sign the final document. When people sign off on documents, they tend to perform a more thorough review of the materials.

Once you have gathered all the stakeholder requirements, I recommend that you pull them all together in one meeting to discuss your findings. This approach indicates that project success is a team endeavor, and while each stakeholder has a different interest or point of view on the project, that they'll achieve success as a team. In addition, by reviewing and discussing the requirements as a team, you may flesh out additional requirements and identify and address any conflicts among stakeholders.

Identifying Course Content Needs

At this point in the process, you've identified your stakeholders and you have an understanding of their success criteria. Now it's time to identify the content that will solve your business problems and enable your employees to meet the organization's expectations. You probably have an idea of the topics that need to be covered, but where do you go from there? How can you validate the content will address your primary audience's needs? Two key areas must be considered when identifying your content needs: understanding your employees' knowledge, skills, and attitudes and understanding the strengths and weaknesses of the existing training programs.

Stakeholder Requirements Questions to Ask

When gathering your stakeholder requirements, make sure that you dive deeper than merely a laundry list of requirements. The following questions can help you gain additional insights into what your stakeholders bring to the table.

- What requirements must be met for this project to be a success?
- When do the stakeholders need to be involved?
- What are their goals for the initiative?
- How will they define success?
- What risk factors do they foresee?
- How often and in what form would they like project communications that help them know their requirements are being met?
- What type of communications do they need so you gain their buy-in and adoption of the program?
- When do they believe the communications to their team should begin, and how often do they want to hear from you?

Identifying Your Primary Audience's Needs

The first area you must consider in identifying your content needs for mobile learning is your primary audience's knowledge, skills, and attitudes (KSAs). The two points here are (1) understanding your target audience's current state and (2) comparing that to the KSAs you believe are necessary to solve your business problem. This does not mean you need an in-depth gap analysis at this point. Your understanding of the gap between the employees' current KSAs and your desired future KSAs will give your Learning and Development team a head start and point them in the right direction for when they perform the detailed gap analysis. Remember, you are not yet designating the actual learning objectives for your program; let the training professionals take the lead on determining them.

Gap analysis The detailed analysis of your employees' current knowledge, skills, and abilities (KSAs) compared to the desired posttraining KSAs. The gap or difference between the two will determine the training content needed as well as the specific learning objectives for your mobile learning solution.

One technique for identifying your learners' content needs is by shadowing them on the job. This allows you to experience

TRICKS OF THE TRADE

Identify the Low-Hanging Fruit

It's easy to overdesign a mobile learning solution. A good starting point when scoping out your solution is to identify the content or information that will easily address the problem. For example, if you are providing a solution to increase the speed at which technicians perform repairs, identify the top 10 repairs that they make and start with those. You'll have a quick win, gain feedback on the results, see where you need to adjust the training, and see results faster. You can always address some of the less common repairs later.

firsthand their struggles and how they approach those challenges. You'll be able to see if and how they currently use the tools provided for them and identify training gaps. You may also notice processes that need modification for the desired future state to become reality.

Another technique for identifying the needed content is to ask your employees. Yes, this is another opportunity where surveys can come in handy, or better yet, you could interview the employees to gain additional context into where they are struggling or even how the problem could be addressed with your mobile solution.

Once you have identified the gap between the current and future KSAs, you'll be able to generate a list of topics that the training must address and how these topics will support your performance objectives. Now you can look at your existing training to identify if the content already exists or how the training material is being presented.

Analyze Your Existing Content

Your company has likely already spent a good amount of time and money on creating training programs for your target audience on the subject matter you'll address in your mobile solution. Take the time to compare

Tips for Mobile Learning Content

When defining your content, keep the following two tips in mind:

1. Design content that fits how the employees use mobile devices. Remember: we use mobile devices in short bursts. Your content should be broken down into bite-sized nuggets.
2. Keep your content simple. Don't provide extraneous information. Only give your employees the facts they need. Then consider providing links to additional content, if helpful.

The Training Mindset

It's important to understand your Learning and Development team's mindset. Typically with instructor-led and e-learning courses, the training spends a good amount of time adding context to the learning experience. Trainers do this by using stories, scenarios, animation, and pictures. In addition, your trainers probably explain what they will teach, why it's important to the audience, what content they will cover, when to apply the content, and the steps involved. At the end of each training segment, these points are typically summarized for the learner.

While this is great for formal training programs, mobile learning is different. In mobile learning, you have a short time to cover your material, and often, the context is already present as the learners are learning within the context of their jobs. If your training team is new to mobile learning, you'll need to reinforce those points often until they catch onto the mobile mindset.

your training programs to your identified content needs. Identify the strengths and weaknesses of those training programs, focusing on how your mobile solution can augment that experience to achieve a greater level of understanding or performance improvement.

Points to consider when speaking to your training and development team about the existing training and supporting content include:

- What content is already provided to your learners, and what are the delivery methods?
- Is the program a blended experience, or is it formal?
- Is the training program an off-the-shelf course or custom built for your company goals and objectives?
- What is your primary audience's reaction to existing training?
- Do you provide all the information at once, where there is a high likelihood they won't retain the key points?
- What gaps exist in the current content that you must address to prepare your employees to be successful in the future?
- Does the existing training program encourage collaboration and informal learning opportunities?
- Why do you believe the existing training courses are ineffective?
- Where do opportunities exist to enhance the current program with a mobile delivery option?

MOBILE LEARNING CONTENT QUESTIONS

Here are some questions to answer as you define your mobile learning content needs:

- What content do your employees need to increase their KSAs?
- When do they need this content, and what will they be doing when they need it?
- What is the smallest amount of information you can provide to them to increase their KSAs?
- Is the content broken down into short learning moments to support how they use their mobile devices?
- Does this content already exist in another training format?
- How does your mobile learning content blend or augment other training programs?
- What opportunities exist for providing mobile learning tools to support your staff?
- Does the content encourage collaboration and knowledge sharing among employees?

Documenting Environmental and Situational Considerations

Once you have a handle on your mobile learning content needs, your next step should be to document any environmental and situational considerations that can impact your mobile learning solution. If you're working with your Learning and Development team on documenting your requirements, this may be new territory for them. Keep in mind that most formal training programs such as ILT or e-learning are designed for specific delivery environments. Teams typically don't need to consider how the environment and/or the situation in which the employee interacts with the training impacts the overall solution. So, you may be wondering, what kind of environmental considerations could have an impact on the design of the program? Let's look at three examples.

Imagine for a moment that you'll be providing mobile learning opportunities to your retail sales representatives. You want to maximize some of their free time to learn more about the products you sell. Sounds great, but how will you deal with the customers' impressions of your employees walking around "playing" with their phones or tablets? That

> **EXPERIENCE IT FIRSTHAND**
>
> MISTAKE PROOFING
>
> When reviewing and documenting environmental and situational considerations for mobile learning, don't make those judgments from your desk. This is one time you need to leave your office and go into the field. Put yourself in the shoes of your primary audience, experience their environment and situation firsthand, and perform your own observations. In addition to personal experience, ask your employees for their input into what situational or environmental concerns they have when using mobile learning. You will gain a tremendous amount of information that will affect what you design as well as the success and adoption rate of your mobile learning solution.

situation should be addressed prior to rolling out the mobile learning solution. In this example, it probably does not impact the design of the mobile learning experience, but you'll need to communicate to your employees how to handle the situation. One possible way to address that situation is to ask your employees to wear an "I'm Training" button (visible to customers) while they are on the floor and actively engaged with the mobile learning.

Next, let's look at an example that impacts your mobile learning design. Imagine that you were designing a mobile learning solution to aid automechanics as they made repairs. You must take into consideration that when doing repairs, their hands are not available to access a mobile device. So how, you ask, will they be able to use a mobile device for support? One answer: they could use special goggles and speech recognition, as in the BMW augmented reality example we discussed in Chapter 4.

Let's look at another example for the design of your mechanics' training. You could blend your mobile learning solution with formal training. After the employees have completed the formal part of the course, you could send them a text message that presents them with a problem and asks them to document how they would approach the problem. Since your employees are always on the go, you could provide them with speech recognition technology to document their thoughts versus having them type on the mobile device or wait until they got to a computer.

Here are more points to consider:

- Is the environment noisy? Is it safe to wear earbuds, or would they put the employees at risk? If they'll be at risk, then audio or video shouldn't be part of your design.

- Do your employees always have access to a Wi-Fi network? If not, you should design a solution that caches the data up front so they won't need to depend on a Wi-Fi connection to be able to use the mobile learning.

Cache To store a copy of data retrieved from a website or mobile application on your mobile phone. When data is cached to your smartphone, you don't need to have an Internet connection to access the data. Your smartphone will pull up the data from the phone's memory. An example is a weather app for your smartphone. If you don't have Internet access, when you open the app, it displays the weather forecast from the last time you accessed the app.

If you'll be sending text messages to a mobile audience, are you at risk that they'll view them while driving? If so, consider adding a prohibition to your acceptable use policy.

Depending on the type and intent of mobile learning experience you are creating, you may find that you do not have any significant environmental considerations or constraints. I recommend that you take time early in the process and think about when and where your employees will primarily access the content or use the solution. Ask yourself and your learners what environmental or situational factors could impact the solution's use or effectiveness. If you identify some impediments, document them now so you can work with your team during the design phase to ensure that you have covered all your bases and you have created acceptable use policies.

Planning for the Effect on Employees of the Mobile Learning Initiative

When managing a mobile learning initiative, it's imperative that you address how this new learning platform will change your organization and you prepare for that change early in the process. Use a three-pronged approach that includes using champions, creating awareness throughout the organization, and providing support.

Identifying Champions

One way you can manage this change within your organization is to

PREPARING YOUR CHAMPIONS

Champions can go a long way in helping your employees accept the changes that will take place and aid in the buy-in and adoption of your mobile learning solution. However, don't rely on the chance that they are prepared to take on the role. Make sure you provide them with guidance on the importance of their role. Talk through ways they can communicate their message to employees. Remind them to speak directly to the employee benefits and not only the business perspective. Last, set expectations with them on communicating their insights from the field back to you.

identify champions who will increase program buy-in. Think of your champions as program evangelists whose primary goal is to get others excited about the opportunities and help them understand the real-life benefits. When identifying your champions, look to two groups: leadership and peers.

First let's address your leadership champions. Your executive team and managers need to walk the walk and talk the talk. We all realize that if employees hear from the top that mobile learning is just a fad or lacks value, then the employees are not going to buy in to the new approach.

Now let's look at peer champions. They should be people within your target, or primary, audience peer group. Ask yourself who within your primary audience has influence with the other employees. People are deeply motivated by their peer group and having coworkers who perform the same role and experience the same challenges to promote the mobile learning program. Peer enthusiasm can go a long way in generating buy-in throughout the organization. Peer champions are able to speak to how the mobile learning efforts have simplified their work or made on impact on how they learned new skills.

Here are a couple of examples. Let's say your mobile learning solution is geared toward extending the learning process. Your champions should speak to how this process helped them remember and apply key points. Let's say you're rolling out a new mobile learning program designed to help your sales team overcome customer objections. Your champion should be able to give concrete examples of how this approach helped him or her in a real-life situation. How did the mobile learning prepare that champion to address the client's concerns and to close the deal? Let's look at one more example. Let's say you design a mobile learning project to provide performance support in the field. Your champions would speak to how the mobile performance support easily walked them through the steps and how they didn't have to remember or guess what the next step in the process was.

Increasing Awareness

Unfortunately, in many cases when a new learning program is released, the company hasn't put a solid communication plan in place. This should not be the case with mobile learning. You'll be changing how your employees think about learning and when learning takes place. You need to let your employees know what is coming, and you need to start early in the process. At three key times, you should actively build awareness for your mobile learning initiative: before the event, at the launch, and periodically thereafter.

> **Two Sides of the Communication Coin**
>
> CAUTION
>
> As managers, we're wired to speak to the business perspective when we talk about a new project. After all, that's the key message when we are speaking to executives and senior leadership. However, from your employees' perspective, the most important point to cover is "what's it in it for me." Explain to them how this new mobile learning and/or job support will make their jobs easier. Sure, it's important they understand the business side of the solution, but focusing on the employees' perspective will go a long way to increasing buy-in and adoption rates for the mobile learning solution and increasing the odds that you'll achieve your desired business results.

First, prior to the event, communicate to your employees what is coming down the pipeline. Don't wait until you roll out the new solution. By waiting, you lose the opportunity to generate buzz in the organization about this new and exciting approach and to give employees time to adjust to the new idea. There's nothing worse for an employee than to come into work one day and find out through the grapevine that things have changed. Remember to get people excited about the new opportunities ahead of time and to give them some tangible examples of how they'll benefit from this approach.

Next, you should communicate with your team when you roll out the program. I normally recommend that an executive or senior manager perform this communication. Let the employees know why the business has invested in mobile learning, what the expectations are for the program, and how it will benefit the employees. All too often, we focus on the business aspect of the change when communicating with our staffs. If this solution is going to simplify their jobs, sell them on the benefits of the change, not on how the company will become more profitable. You should also take this opportunity to explain any mobile device policies that you have created and where employees should go for support on the program.

Last, communicate with your team on the success and progress of the program. Is it working? What results has the company experienced? What benefits are the employees identifying? What changes or adjustments do you anticipate making to the program? Your employees are key ingredients to the success of your mobile learning solution. It's important to let them know how the program is working out.

SMART MANAGING

KEEP YOUR EMPLOYEES IN THE LOOP

Unfortunately, many times a communication plan is an afterthought. The mobile learning solution is about ready to deploy and someone brings up in a meeting: "How will we communicate it to the employees?" Don't allow yourself or your team to fall into that trap. Involve your employees throughout the process. Assign someone to be in charge of the communication plan, and ensure he or she has a solid plan for when and what will be communicated to your employees. At a minimum, the plan should include communication efforts that share information before the mobile learning solution hits the field, when you roll it out, and while the program is running. Remember, your employees are a key ingredient in the success of your mobile learning initiative, and you should do everything possible gain their buy-in and get them excited about this opportunity.

Providing Support

Ideally, your mobile learning solution is designed to be easy to use and even intuitive. Does that mean you won't need to provide support? Think again. When planning for a mobile learning solution, consider what type of training your employees will need to ensure the program's success. Will you need to provide a session on how to use and care for the devices themselves? Will you need to explain the acceptable use policy? Will you need to educate them on how to use the program?

Even if your audience is comfortable with mobile devices, a small percentage will need some basic training on the device. For example, a reasonably savvy friend of mine thought that when he put his iPhone to sleep, that was the same as turning it off. He was wrong and said, "Who knew? Guess I haven't turned the phone off since I got it." It's a good idea to provide at least a job aid on the device to ensure everyone has the same level of knowledge.

Don't Be Naive

Like any other technology-driven solution, there will be problems or challenges for your primary audience when you roll out your mobile learning solution. Even with the best planning, some people will experience issues or struggle with the technology. Make sure that you plan for this by providing them with the support they need to be successful. This may include a help desk they can call or experts in the field your employees can reach out to for support. Remember, if they are frustrated and can't get help, they won't use the mobile solution.

In addition to training employees on the device, this is a good opportunity to cover acceptable use policies. For example, if you are providing your employees with iPads, what is your policy on downloading personal apps or a music playlist? Do you legally need to communicate with them that they cannot use the device to check e-mail or respond to text messages while operating a motorized vehicle such as a car or forklift?

Depending on the type of mobile learning solution you deploy, you may want to provide an awareness piece on how to use the program. You could record a video, hold a webinar, create a job aid, or have an expert in the field share the information. Don't overthink this solution. Just as your mobile learning should be designed to keep it simple, so should any training that you provide to your employees on its use.

In addition to anticipating your employees' training needs, you must also address where they'll go to get answers to their questions on the program and when things go wrong. Even with the best-developed mobile learning solution, people will have questions. Maybe the device locks up; maybe they can't connect to the Wi-Fi; maybe they are experiencing an

QUESTIONS TO CONSIDER ON PROVIDING SUPPORT

While mobile learning programs should be easy to use and require minimal training, here are eight questions you should address when planning for the level and type of support your audience will need.

1. What device-specific topics do you need to provide training for?
2. How will you provide employees with device training (if needed)?
3. How will you educate employees on the company's acceptable use policies?
4. What do you need to train for that is specific to the mobile learning program?
5. How will you train employees on the mobile learning program?
6. Whom do employees contact when they experience problems?
7. How will you track support calls and issues?
8. What types of reports do you need regarding support calls and how often?

error message or the phone isn't working the way they think it should. It's critical to provide support personnel to help your employees work through these problems. On-site experts as well as a help desk are great ways to provide this level of support.

Manager's Checklist for Chapter 6

- ☑ Stakeholders are those individuals who have a "stake" in the success of the project. Remember to include yourself.
- ☑ Meet personally with all stakeholders to understand their requirements for a successful project. This includes their goals for success, views on mobile learning, potential risks, limitations they see with using mobile devices, and their communication preferences during the project.
- ☑ Document all environmental considerations or constraints that will impact the design of your mobile learning experience or require you to create acceptable use policies.
- ☑ Mobile learning content should be based on increasing the knowledge, skills, and attitudes (KSAs) of your employees so they can address your business problem.

- ☑ Mobile learning content should be short, concise, targeted, and simple.
- ☑ When dealing with the change associated with mobile learning projects, you must proactively communicate "what's in it for me" to your employees using champions and awareness communications before, during, and after you have rolled out your mobile learning solution.
- ☑ Anticipate and plan for your employees' training needs regarding the mobile device, the mobile learning solution, and the company's acceptable use policies regarding mobile devices.
- ☑ Provide a support desk that your employees can contact when things go wrong or when they have questions on the mobile learning program.

Technical Considerations of Mobile Learning

I say many times in this book that mobile learning isn't about the technology; however, as a manager, there are a few technical points you should be aware of. In this chapter, we home in on the major points of who provides the device, security risks, technical considerations for developing apps for mobile platforms, and tracking your mobile learning solution. This chapter provides you with an overview in each of these areas so you can have an educated conversation with IT as well as your development team and/or vendor.

One disclaimer I must make about the content of this chapter: technology changes quickly in mobile learning. What is true today while I'm writing this may not be 100 percent accurate when you read this chapter.

Personal vs. Company-Provided Devices

You've decided that mobile learning is a good fit for your company. Now you should think about who will provide the devices your employees will use. Should you let your employees use their own devices, or should your company provide them? Sorry to say, there's no clear-cut answer to this question; you'll need to determine the right choice for your mobile learning solution and your company. What I can tell you is that many companies are leaning toward BYOD, especially as it relates to mobile learning. So let's start by looking at some of the pros and cons of BYOD.

Mobile platform The hardware and the software that run the mobile device. Typically, when people refer to a mobile platform, they are referring to the mobile device's operating system. Examples of mobile platforms are Apple iOS, Google Android, BlackBerry OS, and Windows Phone.

BYOD (Bring Your Own Device) Companies allow their employees to use their personal mobile devices for work purposes.

Pros of BYOD

When talking about BYOD, I like to break it down into the company's perspective and the employee's perspective.

Let's start with the company's perspective of cost savings. You're probably thinking that if your employees use their own devices, then you won't have to invest in that technology. True, but the cost savings are potentially more than the cost of the device. You may also save on the voice and data plans. In many cases, employees are happy to use their existing plan as an offset for the increase in convenience this provides them. In some cases, companies supplement the employees' existing data or text plan, but don't foot the entire bill. If the company purchases the devices, then it is also responsible for the cost of the monthly plans.

Another benefit from a company perspective is that it is able to take advantage of new mobile technologies without investing in new devices. Do you remember the last time your company bought you a computer and how long you had to wait to get a new one? Personally, I used a company-provided laptop that was so old that the TSA agents at the airport would comment on its age. I had to appeal to management for a long time to get a new one. Not an uncommon story. Buying computers is costly, and because the technology advancements aren't as fast or dramatic, you can purchase an older one. Not so with mobile technologies. Each year if not sooner there are significant changes to the processors, cell networks, screen sizes and resolution, available sensors, and more. These technology advancements can have a drastic impact on your mobile learning solutions.

Imagine purchasing iPhones for your employees, and within a year, the product is upgraded. Maybe you don't really need the latest and greatest technology, but it doesn't take too long before the technology you purchased is out of date. If you allow BYOD, chances are high that employees will upgrade their devices more quickly than your company. On average, I'd say that mobile phone power users upgrade close to once a year, and some even more often. The rest of us aren't as apt to upgrade as soon as the newest technology hits the street, and on average, it's probably closer to every 18–24 months. Still, this gives you the ability to leverage the new technologies more quickly than most companies are willing to invest in it.

To summarize, by letting employees use their own mobile devices, you can achieve cost savings in three areas: initial investment in the devices, service plans, and upgrading the devices.

These are the most talked-about advantages of BYOD, but there is one more that I'd like to mention: increased accountability for the devices. When employees use their own device, they tend to be more conscious of losing it or letting it sit somewhere unattended.

I have seen employees leave company-provided devices on their desks when they leave the office for the day and not give a second thought to the possibility of someone stealing the device or accessing the information on it. This happens even when company policy clearly states that they should always store the device in secure locations.

Would an employee do that if it were his or her own mobile device? Probably not. They'd either go back to the office to get it or call someone still at the office to verify the phone was still there and ask that person to hold on to it until the next day. If neither option would work, the employee probably would spend the night worrying about it. After all, their personal information is on the device, and they don't want it to fall into the wrong hands or have someone steal it.

People also tend to take better care of a device when it's their own. After all, if it is damaged, the employee would have to spend personal funds to replace or repair it. When it's the company's equipment, the employee knows the company will replace it at no cost to him or her.

Now that we've covered major advantages from the company perspective, let's look at them from the employees' perspective. There are two main benefits to the employees: increased satisfaction and convenience. BYOD allows employees to use a mobile device that's already part of their daily lives and one that they are comfortable using. That means they are usually happier using the device of their choice versus using an unfamiliar device the company chose for them. Not only do the employees not have to learn how to use the company-owned device and deal with different keyboards or screen sizes, but also they have a level of control over their experience. Another way BYOD increases employee satisfaction and convenience is by reducing the number of devices they need to carry with them on a daily basis.

SMART MANAGING

COMPANY-PROVIDED DEVICES MAY DECREASE EMPLOYEE PRODUCTIVITY

People are creatures of habit. We're accustomed to how our mobile devices work, and using them is second nature to most of us. However, think back to when you first got your new smartphone. You had to learn how it worked and get accustomed to the hardware. Not all smartphones are created equal. Screens are different sizes; keyboards are different, and so are the operating systems. For iPhone users, switching to an Android device can be a frustrating experience. Even for Android users, no two Android devices are the same. The Android operating system is open source, which means that the hardware manufacturers can customize the device's hardware and features and how users interact with it. With this in mind, when you ask your employees to learn a new device, chances are you'll lose productivity while they figure out how to do things that they already know well on their own device.

You may also have a long-term productivity loss caused by the hardware, such as a keyboard. If the device your company uses has a small keyboard, someone with large fingers will always be frustrated and error prone when using it.

The Cons of BYOD

Now that we've discussed all the great ways that BYOD can benefit your company and your employees, let's talk about some of the potential negatives or challenges that you should consider. First, let's talk about the downsides that your IT department will probably mention: security risks and support costs. If your company hasn't already adopted a BYOD policy, then your IT department will probably want to educate you on the security risks for the company and the challenge for IT in supporting all the various mobile platforms. Let's first address the IT concerns relating to BYOD.

BYOD Policies

MISTAKE PROOFING

Your IT department may already have a corporate BYOD policy in place. If that's the case, make sure that you understand the policy and how it will affect your mobile learning solution. Also, ensure that your employees understand what the policy means to them. BYOD policies should include, at minimum, information such as approved devices, the level of support the company will provide, security policies, acceptable use policies, ownership of apps and data, identification of any banned apps, and a policy for making sure the information on these devices is returned if the employee leaves the company. If the policies don't exist, talk to IT about developing them and how you can be a player in crafting them to ensure your mobile learning requirements are understood and achievable.

Let's start with the concerns over increased security risks. What is the level of risk if an employee loses his or her phone? Is there confidential information stored on the device? Is it possible that an unauthorized user could access the company network through the device? These are critical concerns that we discuss in the next section, so for now, hold your thoughts on this topic. Realize, however, that allowing your employees to use their own devices opens the door to additional security risks that you must address.

Another challenge associated with BYOD is providing support for all the various devices. How can IT manage and support the variety of devices that use different operating systems? How can you separate the personal use of the mobile device from business use? How can your help desk support devices with which they are unfamiliar? What are the costs associated with this additional level of support? These are all great questions and valid concerns that your IT department must address.

Is there a solution to these challenges? Some companies provide the devices, which limits the scope of the risk. For companies that embrace BYOD, chances are they use now or are looking into mobile device management (MDM) software that addresses these concerns in a BYOD world. Even companies that provide devices may use MDM to automate and simplify managing mobile devices and provide additional layers of device-level security.

One last consideration or challenge that I'd like to bring up regarding BYOD is specifically aimed at you, the manager. If you let your employees

KEY TERM

Mobile device management (MDM) Software or tools that allow your IT department to remotely set up and configure, monitor, and support mobile devices in your organization. For example, by using MDM your IT department can remotely send updates to operating systems, set up individual profiles for access to network resources, remotely support and troubleshoot devices, and enforce device-level security features.

use their own devices, you could increase the scope of your development efforts by creating apps for an audience that uses a variety of mobile platforms. Should you develop a native app or a web app? What are the pros and cons of each approach? We cover that technical concern later in this chapter.

Security Considerations

Now that we've discussed the pros and cons of BYOD, let's look at security. When you think about mobile learning and security, what questions come to mind? Let me share with you a few of the most common questions that people ask me.

- What if my employees lose their phone or tablet?
- What do I do when an employee leaves his or her job and has mobile learning applications loaded onto his or her personal device?
- How do we protect our proprietary company information?
- We plan to develop native mobile learning apps; however, we can't have them available at app stores where everyone, including our competition, can see them.

THE MOBILE LEARNING SECURITY MYTH

With the deployment of any enterprise technology, you need to do your due diligence to secure your proprietary company information. Mobile learning is no different; however, many mobile learning and security experts such as Robert Gadd of OnPoint Digital believe that mobile learning is actually more secure than your eLearning courses. Work with your IT department. Document your security concerns and compare them to IT's concerns. Then create a plan that addresses everyone's security concerns. Don't let the myth or misconception that mobile technologies aren't secure stop you from moving forward with your mobile learning vision.

These are valid questions and concerns, but they can all be adequately addressed. In fact, your company may already have mobile application management software in place that addresses these mobile security issues on a broader scale than only mobile learning.

Mobile Application Management Software (MAM)

I don't want you to get the impression that mobile application management (MAM) software is only about security. Security is, however, one of MAM's primary features. What MAM software does on a broader scale is allow you to distribute and manage the applications on your mobile devices, provide application-level security, and track usage. Now that you understand the big picture of MAM, let's return to those security concerns.

> KEY TERM
>
> **Mobile application management (MAM)** Software that gives your network administrator control over the apps that reside on your employees' mobile devices. MAM software allows you to create your own company app store and publish your content, add application-level security, and track your employees' use of the content. Like any other software program, app functions and capabilities differ from one vendor to another.

Wondering what to do when your employees leave their device unattended, misplace it, or lose it? You have two issues to consider in these scenarios. First is the short-term issue that someone will find the device and access your apps. Through MAM, the authentication process is tracked, and if it fails, the app is disabled. For example, let's say someone has entered the wrong credentials (password) three consecutive times. At that point, MAM software kicks in and disables the app. Knowing this should help you sleep at night. What if your employee loses the device or leaves the company? MAM remotely wipes your company's apps and associated data from the mobile device.

> KEY TERM
>
> **Wipe** A security feature that allows a network administrator to use MAM to erase data from a mobile device. Using MAM only erases company-specific applications, leaving the employee's personal apps and data alone.

Worried about having your enterprise mobile learning apps on the

MAM Doesn't Secure the Physical Device

If you want to have security on the mobile device itself, not only on the apps that you create, then you need to talk with your IT department about Mobile Device Management (MDM) software. MDM allows your IT department to manage and support the various mobile phones that your employees use. One of MDM's functions is to add device-level security. At the device level you can enforce a device PIN number or password, wipe the device clean, remotely lock down the device, perform jailbreak detection, and activate and/or add encryption. Often MDM vendors include some MAM functions; however, they really are two distinct products. MDM controls the device; MAM controls the apps on the device.

public version of Apple's App Store or on Google Play? Of course you are. Seriously, who really wants to have their company's learning programs in a public venue? One of MAM's features allows you to develop your own private app store (including your own branding) and publish your apps to that location. You can even designate which employees should have access to which apps. As I use *apps* here, I'm referring to more than apps. In many cases, you can also publish your other training documents and multimedia files, too.

How does MAM work from a user perspective? To review the list of mobile learning apps, the employee only needs to select your store's icon and select the mobile learning app to download.

It can even be easier than that. Some MAMs allow you to push (download) applications directly onto your learners' devices without requiring them to do anything. This feature is called an over-the-air installation. You can also push out app updates over the air. MAM not only adds the security, but also simplifies your publishing process.

Using MAM to Measure Your Program's Success

MAM programs can track a variety of usage statistics that help you determine if your mobile learning solution is effective. For example, you can track who has downloaded each of your mobile learning apps, when and how many times a specific individual has accessed any app, on which platforms this employee is using the app, and a variety of other usage data. If your company already uses MAM software, make sure you understand what data the program allows you to track and report on. That will help you evaluate your mobile learning solution.

Mobile Platforms

Up to this point, we have discussed the pros and cons of BYOD, prepared you for conversations and concerns your IT department may have, and hopefully, assured you that your security concerns can be addressed with MDM and MAM software. Now let's discuss mobile platforms.

It's no surprise that the mobile platform market is constantly evolving. In the not-so-distant past, BlackBerry was the choice of smartphone users. However, the market has changed drastically in the last few years, and chances are that about two thirds of your employees are using either a Google Android or an Apple iOS platform. For the remaining employees, they probably use BlackBerry OS, Samsung Bada, or Windows iPhone.

Why should you care about the platforms? Two main reasons: security and apps development considerations.

Security

I know we've already covered security, but you should be aware that each mobile platform has its own security built into the device. Easily the most secure platform for enterprise use is the BlackBerry OS. Even today, President Obama uses a BlackBerry because of the high levels of security the platform offers. As we discussed earlier, however, probably only a small percentage of your employees (and most of them probably at the executive level) use this platform.

While the Google Android is one of the two market leaders, that platform today is the least secure. Your typical employee may not give much thought to security when buying a mobile device. However, your IT department will be concerned about the security of various devices at the platform level.

Native vs. Web Apps Considerations

The next issue you should understand is the technical considerations that the mobile platform brings to the table with regard to developing mobile applications. I find that when I speak to managers about mobile learning, their first inclination is to create a native mobile learning app that runs on a variety of mobile platforms and devices. Before you jump into creating a native app, there are a few technical considerations you should be aware of to ensure that it's the best decision for you.

Earlier in the book, we touched on two approaches to creating an app: native and web. You can create a native app that's developed specifically for each platform and stores all the application data on the device. You could also develop a web application that will run on multiple platforms with a single set of code, but everything is accessed from the mobile browser and the Internet. Managers often think "an app is an app is an app" and do not understand the differences among the options. Take a moment and review Table 7-1 where I summarize some of the basic technical differences between a native app and a web app.

When would you want to use a native app versus a web app? There are by no means hard-and-fast rules, but here are a few thoughts. A

> CAUTION
>
> **ANDROID PLATFORM DEVELOPMENT CONSIDERATIONS**
>
> Many people are excited about the flexibility that the Android open source operating system provides. However, it does present some challenges that your team must proactively address and plan for when designing and developing native mobile applications. Don't assume that your native apps will automatically work on all devices. Challenges include:
>
> - Each manufacturer controls which version of the Android platform its devices run on.
> - There is no standardization among Android devices. The manufacturer can add as well as disable features. For example, Amazon Kindle Fire uses the Android OS; however, the web audio feature has been disabled. The manufacturer also controls which buttons on the device do what functions. What this means to your development team is that they can't rely on the function a particular button may perform.
> - Updates to the OS are released by the carriers (i.e., Verizon, Sprint, Motorola, AT&T). It is common for phone owners to be running on older software. If your company uses MDM, this may not be an issue because MDM can push the updates to your employees' phones.

native app is a good choice when you want to use the unique features. It's also the right choice when you need a rich user experience and interactivity such as in a mobile learning game. A web app may be the right choice if your content is primarily text based and doesn't require a lot of user interactivity. If your content is constantly changing, then a web app might be the right avenue to take, since updates are available immediately. A web app is also a good choice when you need your apps to be

Attribute	Native Apps	Web Apps
Platform	Designed specifically for each platform	Runs on multiple platforms
Development tools	Platform-specific software development kits and integrated development environments	One set of code using common web development tools such as HTML5, CSS, and JavaScript
How accessed	Downloaded and stored on the device	Mobile browser
Internet connection	Not needed to access content	Required
Content updates	Requires updating app and downloading to the device	Immediately available upon publishing
Accessing unique device features	Full access	Limited access
Performance	High level of performance	Lower performance based on network connectivity
Rich user experience and interactivity	Very high	Low

Table 7-1. Native app vs. web app technical considerations

accessible from a wide variety of devices since you'll only need to develop one set of code.

Aren't All Browsers the Same?

CAUTION

Most of us nontechies would assume that a browser works the same if you are on a mobile device or if you are accessing the Internet from your laptop. In reality, mobile browsers work much differently than computer-based Internet browsers. For example, if you are accessing a web app from your mobile device that contains an audio file, the web app will not auto play. You'll need to click on a link to hear the audio file. In some cases this will also open a new window. Each mobile browser is different, so your team must be aware of the limitations of the various mobile browsers and how those differences will impact your solution's design.

Hybrid Apps

There is actually a third app choice to discuss with your development team; it's called a hybrid app. The hybrid app combines some of the best attributes of native apps and web apps in one solution. In a nutshell, your content is developed as a web app, then a native app wrapper is placed around the app that lets it communicate directly with the mobile platform. The hybrid app is downloaded directly to your employees' devices, and the learning content is cached to the device. This eliminates the need for an active Internet connection and improves performance compared to a web app. As a result, you can provide the rich user experience and speed of a native app, while benefiting from the flexibility of supporting multiple platforms and the ease of content update of a mobile app.

A hybrid app is not always the best choice. For example, if your goal is to create a highly interactive mobile learning game, then a native app is probably the best option. If your content is highly text based without the need for a rich user experience, then a web app is probably your best option. At the end of the day, your app choice should be determined based on a solid list of requirements and the option that best fits those requirements, combined with your budget and capabilities.

Tracking Your Mobile Learning Experience

Typically, when people start talking about tracking learning events, their first thought goes to using their existing learning management system (LMS). Since companies have already invested in LMS technology, the first assumption is to add your mobile learning to your existing LMS. However, before you jump on that bandwagon, make sure you understand your LMS' ability to support mobile learning. The LMS

Sharable Content Object Reference Model (SCORM) A set of standards and technical specifications that define how your eLearning courses will communicate with the LMS. The premise behind SCORM is that any eLearning content will play on any industry standard LMS. SCORM was developed in 2000 by the Advanced Learning Directive (ADL) of the U.S. government and has three versions: 1.1, 1.2, and 2004.

industry has been making gains in this area. However, you should realize that it's not only having an LMS that can support mobile learning, but also understanding SCORM and the limitations this standard brings to the table for mobile learning.

Limitations of an LMS and SCORM for Mobile Learning

SCORM is the industry standard when it comes to how your eLearning and LMS communicate. SCORM also determines the data you can track. Your LMS is only going to track what takes place within the course, and even then, you are limited to the data you can store on the LMS. This would include information such as if an employee has completed a course, the total amount of time spent in the course, the last score on an assessment, and if he or she passed or failed the course. Is this data meaningful to you with regard to your mobile learning solution? Probably not.

There are a few other limitations of an LMS that you should be aware of. First, your training must be developed to play within a web browser. In most instances, you must have an active Internet connection. Next, the only time data is saved to the LMS is when your employee closes the course. Last, the learning content must be taken from within the LMS. What does this mean for your mobile learning solutions? You won't be able to track your mobile apps, performance support, immersive simulations, or games using the SCORM standard and your LMS.

Does this mean that we are constrained to this old world of LMS and SCORM? Not at all. Recently, there has been a radical change: it's called the Tin Can API.

Tin Can and Learning Records Store

With the evolution of mobile learning experiences, ADL released a new version of SCORM called the Tin Can API that revolutionizes what and how we can track our mobile learning solutions. Using the Tin Can API, we can track any learning action from any device and overcome SCORM's limita-

> **Tin Can API** A flexible, new standard for eLearning and mobile learning that allows you to track any learning experience from any device and save the data to a learning record store. Also referred to as the *Experience API* or *Next Generation SCORM*.
>
> **KEY TERM**

KEY TERM

Learning Record Store (LRS) A data repository where all the learning records are saved using the Tin Can API. The data stored in the learning record store can be accessed by an LMS or by a reporting tool, or it can be integrated with other internal systems.

tions. For example, you can now track any actions or experiences your employees have within mobile apps, games, simulations, websites, and even on the job. You can store multiple records for an employee. And you can access the content from any location; you're no longer limited to SCORM.

TOOLS

STRUCTURE OF TIN CAN STATEMENTS

If you can dream of an action you'd like to gain information on about your mobile learning solutions, the Tin Can API should allow you to gather it. While there is a lot of technical information about Tin Can statements, from a manager's perspective, let's focus on the high-level statement structure of noun, verb, object.

Any action that can be framed in that context can be saved to the learning record store. If your employees have successfully completed a portion of a game, you can track that. If your employees are watching videos on their mobile phones, you can track that experience as well. The possibilities are not limited to mobile learning, any learning experience can be tracked using the Tin Can API and saved to the LRS.

The Tin Can API will save your employees' actions in a learning records store (LRS), which can be accessed by your LMS and other reporting tools. While Tin Can API has only been out a short time, some of the mobile learning platforms as well as LMS vendors in the market are already adding an LRS to their systems to support the Tin Can API. They are enhancing their current products to take advantage of this new API and the large amount of data we can now save and report on regarding our employees' learning experiences, not only those found in an LMS. This is a game changer for tracking mobile learning. After years of being constrained by SCORM, there is finally a light at the end of the tunnel.

> **EDUCATE YOURSELF ON TIN CAN API**
>
> SMART MANAGING
>
> While you don't need to know all the details of the Tin Can API and LRS, I recommend that you educate yourself on this new concept. Gain a basic understanding of its capabilities and explore some of its use cases. You'll thank me when it's time to sit down with your internal teams such as HR, Learning and Development, and IT. Since this is a new concept, there is a good chance that others in your organization may not be up to speed on Tin Can. In that case, they may try to limit you to previous versions of SCORM, the capabilities of your existing LMS, or suggest using a technology such as web services to save data about your learning experiences.

Manager's Checklist for Chapter 7

- ☑ BYOD is one of the hottest trends in business today. Ensure that both you and your employees understand your company's BYOD policies.
- ☑ There are no standards when it comes to the various mobile platforms. You need to account for varied screen sizes, processor speeds, and input mechanisms as well as the various operating systems.
- ☑ Using mobile device management (MDM) and mobile application management (MAM) software are one approach to addressing security concerns with mobile technologies.
- ☑ Determining the right type of app to create (native, web, or hybrid) should be based on your specific requirements.
- ☑ Tracking abilities for mobile learning are endless using the Tin Can API and learning record store.

Build vs. Buy Considerations

There comes a time in almost any type of project that you must decide if you'll build the solution with your in-house resources or if you'll partner with an outside vendor. In this chapter, I help you make that determination. We start our journey by gaining an understanding of who else in your company has already gone down the path of leveraging mobile technologies, and we give you an understanding of what they may have learned in the process. We discuss the roles that make up a mobile learning development team as well as some of the key skills needed to succeed.

Once you have an understanding of your company's internal capabilities, we discuss the reasons you may want to work with vendors, and we give you some questions to ask potential vendors. There isn't a right or wrong answer to whether a vendor is the right solution, but after reading this chapter, you should be ready to decide the right approach for your project. Let's start by identifying who in your company is getting innovative with mobile technologies and what you can learn from their experiences.

Who Owns the Innovation Cell in Your Company?

To get you started with the build-or-buy decision, I recommend you seek out others who have already gone down the path you are about to embark on. Has a particular group in your company taken the lead when it comes to innovation and mobile technologies? What key lessons have they learned? Do they have technology in place that your team could leverage? What advice can they offer you? Depending on the type of company you work for, you may have a business unit that works with other business units to supply innovative business solutions. You may even have a person with a title like Head of Innovation. If so, he or she is a great starting point. For the majority of us, it won't be that easy to find the group that has taken the lead in innovating with mobile technologies. Let me help you get started by pointing out a few business units you should investigate and how one of them may be leveraging mobile technologies to add value to your business.

TOOLS

QUESTIONS TO ASK

Once you have identified the group that is innovating with mobile technologies in your company, you'll want to find a manager or other key stakeholder you can speak with about what the group has learned about implementing mobile technologies. To help you get the conversation rolling, I've listed 12 key questions to ask.

1. How is that group using mobile technologies?
2. What successes have they had?
3. What failures did they experience, and how did they overcome them?
4. What are the key lessons they learned?
5. If they had to do the project again, what would they do differently?
6. Did they develop the project in-house, or did they work with a vendor?
7. What process did they use to choose a vendor, and what would they change if they had to choose one again?
8. What vendor did they use?
9. What tips do they have for successfully partnering with that vendor?
10. What internal challenges did they encounter, and what tips can they offer for successfully navigating those waters?
11. What resources could they provide to you and your team to increase the success of your mobile learning initiative?
12. What technology is available in-house to support a mobile technology deployment?

Human Resources

For many managers, the first group that comes to mind when it comes to innovating with mobile learning is Human Resources. After all, they probably own the learning space in your company. While that may be the case, chances are high that when you speak to them, you'll find they aren't the primary innovators.

Why do I say that? With this new approach to learning, there is a significant investment to be made in mobile learning research and development and the technologies that support it. In many cases, the HR department doesn't have the approved budget to invest on its own, so another group ends up taking the lead. While HR will be a key partner and stakeholder in your project, chances are HR isn't going to take the lead with this new technology. That said, you should still speak to the manager of the Learning and Development team (who probably reports to the HR director) and find out what experience that team has with designing, developing, deploying, and supporting mobile learning.

HR Can Partner with You on Your Decision

While you may find that the HR group isn't necessarily the innovator in the mobile learning space, it most likely will want to be involved in determining if you should build the solution in-house or work with a vendor. So HR is one of the first groups you should speak with. Here are eight questions to get that conversation started with HR.

1. What is HR's philosophy regarding using mobile technologies to deliver training and performance support? How can HR assist us in bringing our mobile vision to life?
2. Does HR have experience in designing and developing mobile learning? If so, how many programs has it worked on, and what types of mobile experiences did the HR team create?
3. What type of development process does the team use? (We get into this subject and the importance of this question later in the book.)
4. Is your company's learning management system (LMS) mobile friendly? If so, what types of experiences does it support?
5. Has HR deployed custom content or only third-party content to the LMS?
6. Does the company have a mobile learning platform in place? If so, what types of learning experiences does it support, and what type of content creation tools does it provide?
7. What type of professional development has the HR team gone through to prepare them for mobile learning?
8. How can HR envision partnering with us? (This question should cover both the short term regarding making the best decision on build-or-buy as well as throughout your mobile learning journey.)

Mobile learning platform The software and/or hardware to create, publish, and manage a variety of mobile learning experiences. This can include content creation tools, security solutions, packaging and deploying the mobile learning content, and tracking capabilities.

Marketing Group

Another group you should speak with is Marketing. Why would Marketing be interested in taking the lead with mobile technologies? Think about it. This group is charged with understanding the customers' wants and needs as well as owning the brand. Mobile technologies provide a great medium to support that mission. Maybe Marketing doesn't classify it as mobile learning, but chances are it has already invested in and used mobile technologies as a way to increase brand awareness and gain information about current and potential customers.

A few ways in which Marketing may be using mobile technologies include SMS messages, mobile websites, advergames, QR codes, mobile apps, video, and even augmented reality combined with print materials. While possibly not the first group in your company that would come to mind in taking the lead, there is a high likelihood that Marketing has a vested interest in mobile and could be the group that owns the organization's mobile innovation center.

Advergame A video game that includes advertising around a particular company, brand, or product. Companies use advergames in two main ways. The first way is to sponsor a popular video game. While this may include sponsor ads, ideally the company, brand, or product is embedded in the actual game experience.

The second way, which has become popular, is to invest in creating a complete game. The games are distributed through the company's website or through the app stores. For example, the M&M Brand Chocolate Factory game is distributed via iTunes. In this game, the players must save the M&M characters from being dropped into hot milk chocolate where they would melt.

Sales

For many organizations, the Sales group is the company lifeline, so that could be the group that has already invested in innovating with mobile technologies. After all, this group has probably been using personal mobile devices to increase sales effectiveness long before your company had a BYOD policy or before other groups considered using mobile technologies. Chances are good that Sales has already leveraged innovative ways to incorporate this technology. Sales could be streamlining the sales process with mobile apps, setting up an e-commerce site, using mobile performance support to close sales, or even using mobile learning to educate the sales force on new product enhancements or new products.

IT Group

CAUTION

One group that is probably at the front of your mind is your organization's IT department. After all, we're talking about mobile technology, and technology is what IT is all about. Chances are that IT will be a key stakeholder in your project and a lifeline for ensuring the security of your solution and supporting the integration of data with other internal systems. However, you need to ask yourself the following questions:

- Is the IT department really innovating using the technology or is IT only supporting the technology?
- Is IT willing to innovate with mobile learning?

You and your IT department may have different goals in mind, and mobile technologies may present some unwelcome challenges for your IT department. Don't let any resistance from them stop you from moving forward with your vision, and remember to include IT as a stakeholder on your team.

Customer Service

Yes, you did read that correctly; the heading says Customer Service. We may typically view Customer Service as the call center. Unfortunately, for many users, they immediately think of an unpleasant experience with Customer Service. Sure, for complicated problems, a personal call can be the best solution, but many times it seems there should be a faster, more efficient, less stressful answer to the problem. Well, mobile apps may be the answer. Some Customer Service groups are starting to use mobile technologies and apps as a way to give customers a self-service option for their most common problems. You should investigate the Customer Service group as a potential innovator with mobile technologies and a stakeholder.

Look at All the Possible Groups in Your Company

Human Resources, Customer Service, Sales, Marketing, and IT are not the only candidates for owning your mobile innovation cell. Every company is different, so here are a few additional departments for you to investigate.

- Manufacturing
- Research & Development
- Consulting Services
- Product Development
- Shipping and Receiving

Who Makes Up the Learning and Development Team?

Now that you've identified and spoken to each of the mobile innovators in your company, you're on your way to deciding if you should build your mobile learning solution in-house or if you should partner with a vendor. Your next step is to determine whether you have the resources in-house to design and develop your solution and if the staff has the required skills. In this section we discuss the team members and the skills needed, and how they differ from your e-learning team's skill sets. This discussion should provide you with a complete picture for evaluating if your Learning and Development team can ensure your success.

Experience in e-Learning Isn't Enough

Many managers will look to their Learning and Development group to provide resources, and typically these resources come from an e-learning background. While both mobile learning and e-learning use technology, they are different practices. In addition to a grasp of creating learning experiences, mobile learning requires skill sets that are more typical of IT skills such as web design and application development. Mobile learning also calls for IT development methodologies, which we discuss later in the book. When you evaluate the roles and skills for your mobile learning team, speak with your Learning and Development team, but also talk with IT.

Your typical mobile learning teaming will consist of a:

- Project manager
- Subject matter expert

- Instructional designer
- Interface and usability designer
- Developer
- Tester

STAKEHOLDERS ARE PART OF THE TEAM, TOO

I didn't include stakeholders in the participant list as we discussed their importance early in the book. However, don't take this to mean that they shouldn't play an active role on the project. Meeting their requirements is key to the project's success. The complete Learning and Development team should be aware of the stakeholders and their requirements for a successful project.

Project Manager

Considering you are a manager, I'm sure that you're well versed on the role of a project manager, so we'll keep this discussion short. The key roles for the project manager are to oversee and to coordinate the project activities. This includes assigning resources, ensuring the project vision comes to life, establishing timelines and ensuring they're met, managing project risk, and of course, managing the budget.

SMART

MANAGING

PROJECT MANAGER ENSURES YOUR VISION COMES TO LIFE

One of the key benefits that your project manager provides is taking responsibility for bringing your mobile learning vision to life in a way that provides the maximum business value. We often think of project management as only managing resources, timelines, and budget. When it comes to learning solutions, the project manager is the person who makes sure the instructional intent and the business goals are achieved. Make sure you and your project manager are on the same page when it comes to your vision for the project, what defines a successful project, and the stakeholder success criteria.

Subject Matter Expert

Subject matter experts (SMEs) are sometimes referred to as content experts. Depending on the complexity of your mobile learning project, you may find you need multiple SMEs. As the title says, these people understand the content inside and out that your team will be developing. Use your SMEs during the information and content gathering stage as well as throughout the project. Incorporate them into the review team to ensure the content meets user expectations.

Instructional Designer

Traditionally instructional designers are the people who perform the skill and gap analysis, determine the learning objectives and the assessment method for the course, organize the content, design the learner interactions, and write the content for the learning program.

How does this differ from developing mobile learning? Many of the core skills are the same, but there are a few differences. First, the instructional designers need a strong understanding of information architecture as it relates to mobile devices. Next, they must be active users of mobile devices. If they aren't users, they are missing a key ingredient in determining how to best design content and organize information in the ways a mobile workforce will use it. Instructional designers also need a solid understanding of learning theories as they relate to mobile learning. Last, your instructional designers should understand mobile interface design and usability.

TRICKS OF THE TRADE

A New Mindset Is Needed

For e-learning instructional designers to be successful in designing a mobile learning solution requires a new mindset. If they are still thinking in terms of delivery methods such as e-learning and instructor-led training, they won't be successful. Some people make the switch easily; for others, it's a challenge. The following list describes how the mobile instructional design mindset differs from the traditional instructional design thought process.

- **Length of time spent on learning.** Traditional courses are measured in hours. Mobile learning is measured in minutes.
- **Presenting content.** In traditional courses, content builds on the previous content. In mobile learning, we provide just the right content at the right time, and the learners control what they need to know. Understanding how to organize information in a way that makes it easy for learners to find it is critical to this new mindset.
- **Simple and structured.** In traditional courses, a lot of extra information is provided to the learner. In mobile learning, we provide learners with the smallest amount of information needed to solve the problem.
- **Think "mobile delivery" first.** When designing a blended approach, the instructional designer should first consider what is needed for the mobile format, then work in the other delivery methods.

User Interface and User Experience Designer

For mobile learning solutions, you need someone to be responsible for generating the user interface and designing a system that your learners can easily navigate. When talking about the user interface, it's more than the "look and feel" of the system. Rather, it's how the learners interact with all aspects of the mobile learning solution. This interaction includes the navigational elements, how learners search for information, content layout, and all the other possible system interactions. The best e-learning user interfaces are those that are filled with features and functions that encourage active learning experiences. Not so with mobile learning. Simplicity is key.

With mobile learning interface design, the designers need to understand the devices' limitations (for example, screen size, resolution, bandwidth constraints, small keyboards, small clickable areas) as well as how learners will use the devices (for example, one handed, for short bursts of time, with lots of distractions). The designers must create an environment that facilitates ease of use. To be successful, the designers should understand and apply platform-specific guidelines (published by Apple, Google, Microsoft, and Research in Motion) for user interface design and become familiar with mobile user interface patterns, which are documented solutions to common interface problems.

Developer

The developer is responsible for writing the code and bringing your mobile learning solution to life. The skill sets required for mobile learning development are different from a typical e-learning developer. Many e-learning developers use a software program called Adobe Flash; however, Adobe has decided to release no further updates to Flash. While the product is still a strong player in the e-learning world, this skill set brings no value to the table for mobile learning. The skills required for mobile developers resemble those of an application or web developer. What skill sets do your mobile developers need? Take a look at Table 8-1 where I've broken down some of the more common skills needed by type of app.

Tester and Quality Assurance Specialist

In the e-learning world, this skill is often referred to as quality assurance

Type of App	Skills Needed
Native apps for iOS	Apple SDK using Objective C IDE: Xcode
Native apps for Android	Android SDK using Java IDE: Eclipse
Native apps for Windows Phone 8	.Net for Windows Phone IDE:Visual Studio
Native apps for Blackberry 10	Cascades SDK using C++ IDE: QNX Momentics
Web apps	HTML 5, Java Script, and CSS3
Cross-platform apps	HTML 5, Java Script, and CSS plus cross-platform tools such as Adobe Air, Phonegap, Titanium, or Unity
Augmented reality	Layer, Wikitude, or Junaio

Table 8-1. Common software skills required for mobile developers

(QA). Typically in e-learning, you have a predefined environment and limited constraints, such as an identified operating system, browser, screen resolution and size, and minimum bandwidth. From a testing perspective, only a small number of items need to be checked.

This isn't the case with mobile learning projects, which require an extensive test plan to be generated and tracked. The tester has to determine test cases that document all the possible user actions and the appropriate outcomes. Testers must run through a wide variety of devices and operating systems under various bandwidth constraints and possi-

SMART MANAGING

LESSONS FROM IT ON TESTING

The tester's role on a mobile learning project aligns with the role of an IT application tester. The process used, the generation of use cases, and even the systems used to track and resolve the issues are all IT related rather than Learning and Development related. This is an area where IT may be able to assist your team. Speak to your IT department about what testing resources they could offer for use on your project. If they can't provide you with an experienced tester, then ask if they can mentor or train someone on the process and on an online issue tracking system, and offer best practices and tips for generating use cases.

bly off-line. The testers need to walk in the shoes of the end users and test the program under the environmental and situational constraints the employees will experience. Testers need to be detail oriented, creative in devising tests, and thorough. Their goal is to find malfunctions in the solution. Then the testers document the problem and how to re-create it so the development team can fix and test the resolution.

When Should You Work with a Vendor?

This is an easy decision to make if your company doesn't have an internal Training and Development team. If you do have an internal training team, the decision is more complicated. Why would you work with a vendor? The two main reasons in my opinion? Expertise and technology. While mobile learning has been talked about for many years, it's a relatively new concept when it comes to businesses implementing it as part of their learning strategy. The field is also in a state of rapid flux. New technologies come out every few months, and it's difficult for a Learning and Development team to stay up to date on the latest changes. Let's look at the expertise that vendors can bring to the table.

SMART MANAGING

Benefits of Working with a Vendor

While I focus on expertise and technology as the primary drivers for working with a vendor, these are not the only possible benefits you could derive from partnering with a vendor. Here are a few additional benefits to consider.

- Improved product quality
- Increased efficiency (familiar with development processes)
- Increased flexibility
- Reduce risk
- Decreased time to develop the solution

Expertise

There are three areas in which a vendor's expertise may pay off for you. In this section, we cover two of the three areas: strategy and team skills. The third area, development processes, is addressed in Chapter 10. Let's start with the strategy component.

You have already generated your mobile learning vision, but whom can you work with to confirm the approach? Is there a better, more efficient way to achieve your goals? Normally, you would talk with your Learning and Development department; however, if your team isn't well versed in mobile learning, then they probably won't be able to give you enough help (although they should be involved in the process).

When working with a vendor on strategy, it can assist you to:

- Assess your company's readiness for mobile learning
- Decide which platforms and devices to target
- Determine how mobile learning can best fulfill your business goals
- Validate your mobile learning vision
- Create a plan on exactly what to implement and how to tackle the challenge
- Create an evaluation plan
- Create a maintenance plan

Earlier, we talked about the skills needed for a mobile learning team. Do you have the right resources with the right skill sets? If not, you could work with your existing team to beef up their skills, but that could be a costly and risky proposition. You may need an outside vendor's expertise in the design and development of the solution.

E-LEARNING VENDORS AND MOBILE LEARNING

Many companies have asked their current e-learning vendors to help with a mobile learning initiative. After all, the vendors are already familiar with your company and your systems. A word of caution before you commit to them. e-Learning differs from mobile learning, but some e-learning vendors will tell you they can be your one-stop shop for all your learning needs. They may be able to help, but the reality is that unless they have a lot of experience in mobile learning, you should probably look elsewhere.

Your company needs an expert guide who can bring past lessons learned, understands the tools and development methodologies, stays up to date with the latest advancements, and can help you make informed decisions. Don't be a guinea pig for an e-learning vendor who wants to get into mobile learning.

Technology

While I've emphasized that mobile learning isn't only about the technology, there comes a point when you need to identify a technology to deliver and manage your content. Let's talk about working with vendors who provide mobile learning platforms.

Mobile learning platforms offer you a full-service option for creating, publishing, and managing your mobile learning experiences. In addition, many providers offer stand-alone solutions or the ability to integrate with your existing learning management system or other internal systems. If you're looking to move beyond mobile learning apps, a mobile learning platform may be an answer to providing and supporting a wide variety of mobile experiences to your employees. While each platform has its own specific features, some of the learning experiences a mobile learning platform might support include:

- Alerts and notifications
- Interactive messaging (voice and/or text)
- Mobile web content
- Videos
- Podcasts and RSS feed content
- Documents such as pdf's, PowerPoint, and e-books
- Third-party courseware
- Surveys
- Tests

A few of the industry leaders in the mobile learning platform space include:

- **Achieve Labs:** LearnCast
- **Certpoint Systems:** CERTPOINTVLS Mobile
- **Intuition:** Intuition Mobile
- **KeneXa:** Hot Lava
- **Moving Knowledge:** Moving Knowledge Engine
- **OnPoint Digital:** the CellCast Solution
- **Upside Learning Solutions:** Upside2Go

Developing Long-Term Capacity by Working with a Vendor

While many of you will choose to initially work with a vendor on the design and development of your mobile learning solution, that doesn't mean you're tied to that vendor. When selecting a vendor, speak openly about ways to combine your team with theirs to increase your staff's

Questions to Ask Potential Vendors

If you decide that working with a vendor is the right solution for you and your company, you want to make sure to choose the right one. To help you with your decision-making process, I have listed 13 key questions to ask your candidates.

1. How long have you been operating in the mobile learning market, and how many implementations have you performed?
2. Can you show us examples of your work specific to my organization's mobile learning vision requirements?
3. Do you provide services for companies similar to ours?
4. Can our organization directly contact your references?
5. What type of training and support will you provide on the system?
6. What is unique about your technology specific to our mobile learning vision?
7. What is your implementation plan for the system?
8. How will you partner with us to increase our team's capabilities?
9. How do you manage "scope creep"?
10. What measures will you put in place to ensure our goals are met?
11. How will you involve our team in the development process?
12. What type of support will you provide once the project goes live?
13. Why are you the best partner for our company?

knowledge and abilities. Use the opportunity to grow your internal team's skills and knowledge so that they will be able to update the content as well as create mobile learning experiences in the future.

Don't Forget About Social Networking

While there are many social networking platforms available to enterprises on the market today, such as Yammer and Jive Software, only one social networking app focuses specifically on learning. Float Mobile Learning's Tappestry application runs on both iOS and Android devices and can be configured to track your learners' informal learning. This application lets your learners track and share their informal learning with others in the organization through a social network. Since Tappestry is developed using the Tin Can API, all the learning statements can be saved to a learning records store. As a manager, you can run reports and gain valuable information. You can even set up private groups so your learners can share their learning experiences with others. For more information on Tappestry, go to their website at https://www.tappestryapp.com.

QUESTIONS TO ASK YOUR VENDOR TO INCREASE YOUR TEAM'S CAPABILITIES

- Will you provide customized training to our staff? (This could include training on instructional design, development tools, testing processes, and development methodologies.)
- Can you incorporate our team members into your development team?
- Will you assess our current team's abilities and create a professional development plan for them?
- What kind of investment will you make regarding educating us on current trends and techniques (e.g., webinars, newsletters, blogs, passes to expo halls at conferences)?

Manager's Checklist for Chapter 8

☑ Take the time to seek out others in your company who already use mobile technologies. Talk to them about what worked, what they would do differently, how they approached the build-or-buy decision, what challenges they encountered, and if they have internal resources (people, tools, technology) you can use.

☑ Look beyond the IT and HR departments when identifying who owns the innovation center around mobile technologies.

☑ Mobile learning development teams have different skill sets from e-learning teams. Many of these skill sets fall more in the IT world of application or web development.

☑ Your development team must be well versed not only with the requisite skills, they must also be active users of mobile technologies.

☑ Vendors can provide you with the expertise to bring your mobile learning vision to life, best practices, development processes, and proven team and development processes geared specifically to mobile learning.

☑ Choose your vendor wisely. Designing and developing mobile learning is not the same as e-learning.

☑ Look to your vendor to provide you with a mobile learning platform that your team can use to develop, publish, and manage your mobile learning content, or use the vendor's resources to develop one just for your purposes.

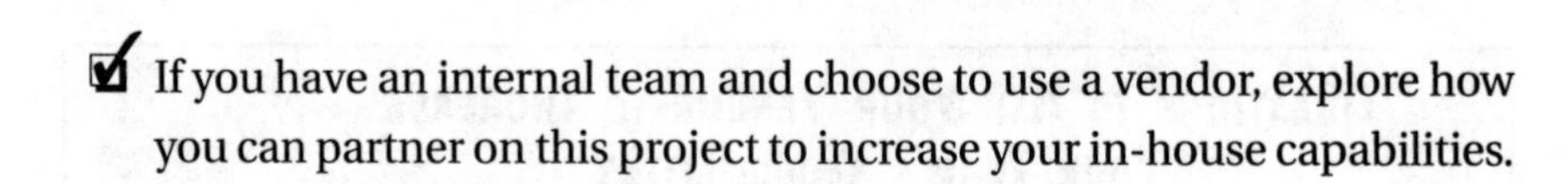

☑ If you have an internal team and choose to use a vendor, explore how you can partner on this project to increase your in-house capabilities.

Gaining Stakeholder Support for Mobile Learning

At this point, you've drafted your mobile learning vision and established a business case, gathered your business unit stakeholder requirements, and thought through whether you have the internal development capabilities to produce your mobile learning solution in-house or if you'll work with a vendor. Now it's time to solidify the support from your stakeholders.

You already should have a good start at gaining stakeholder buy-in to your mobile learning solution since it's grounded on addressing a valid business problem. You have also determined how you'll measure the program's success. However, that is not enough. For your executive stakeholders you must communicate not only the business value of the mobile learning but also how you will minimize program risks and control costs. For the rest of your stakeholders you must ensure their requirements and needs will be met by the solution. Let's start by talking about some tips in gaining executive support.

Selling Mobile Learning to the Executive Team

Your executive team's requirements will differ from the needs of your other stakeholder groups. From an executive perspective, you need to convince them that your solution will have a positive impact and will provide long-term value to the organization. You should also assure

Account for All Your Stakeholders

Remember that the stakeholders of your mobile learning project are not limited to only the executives who have the final authority to move your project forward or cancel it. You also have business unit stakeholders, which can include representatives from HR, IT, Marketing, Compliance, even Legal. You should have already gathered each of their unique project requirements and success criteria. Your employees and/or your customers who use mobile learning are also stakeholders. When gaining buy-in from your stakeholders, have a plan in place that will ensure each group's needs and expectations will be met by your mobile learning solution.

them that you're approaching the project in a way that will minimize risks and costs while maximizing returns. In this section, we talk about how your project benefits in terms that equate to business value, your plan for measuring this impact, and an approach for minimizing risks and development costs.

Benefits Your Company Will Reap

You may be wondering why we're talking about the benefits of mobile learning. After all, we discussed those in Chapter 2. As a refresher, we explored how mobile learning can:

- Transform dead time into productive learning time
- Provide immediate access to critical information that will allow your learners to troubleshoot and make educated decisions
- Improve application of learning content on the job by providing short, targeted, context-relevant educational content
- Provide just-in-time access to learning content
- Encourage employees to participate in training anywhere, anytime
- Encourage the learner to reflect on content
- Put the learner in control of his or her learning experience
- Support formal and informal learning opportunities
- Provide learning content in a way that the Millennials expect
- Increase collaboration among employees such as sharing best practices with user-generated content

When you get right down to it, all of these are ways to increase productivity, improve performance, and support our modern workforce. While these are some of the key benefits of mobile learning and it definitely provides value to your employees and (indirectly) to the organization, these benefits are not focused on the true business impact. This is not what your executive stakeholders want to hear from you. Instead, you must focus on how your mobile learning solution will benefit the organization from a business perspective. When you get right down to it, executives primarily think about the benefits of investing in a project from two perspectives: how to increase revenue opportunities and how to decrease costs.

> **EMPLOYEES WILL NOT WANT TO TAKE MOBILE LEARNING** SMART MANAGING
>
> If your executive stakeholders question whether your employees will embrace the concept of mobile learning and actively participate with mobile learning experiences, you have ammunition to address this question. Talk about the benefits I listed in the beginning of this section such as just-in-time access, increased collaboration and sharing best practices, immediate access to critical information that aids your learners in troubleshooting, and making educated decisions. All of these are really ways that the employees can benefit from mobile learning.
>
> A second point you should make is to explain to the stakeholders that mobile learning is not taking an entire e-learning course on the phone. Once the employees understand how mobile learning will be used, they should embrace the concept.

Start the conversation by explaining the business problem you are solving so the executives have a complete understanding of your current state. Then move into how your company will not only solve the problem through mobile learning but also provide long-term business value to the organization. Focus your business value discussion on increasing revenue or decreasing costs. Depending on the business problem you are solving, some of the business benefits of your solution could:

- Increase sales of your services or products
- Increase the rate at which you produce your product
- Increase the production quality of your product
- Reduce the number of safety incidents
- Reduce the number of customer complaints and increase customer satisfaction
- Reduce the time spent by technicians troubleshooting and fixing routine problems

- Reduce the amount of money spent on formal, face-to-face training programs (cost of creating and producing materials, facility costs, instructor costs, travel expenses)
- Reduce the amount of money spent on e-learning programs (development costs typically range from $10,000 to 35,000 for each hour of content created)
- Reduce the time and travel expenses employees expend on training away from their job

TRICKS OF THE TRADE

QUANTIFY THE BENEFITS

When talking to your executive stakeholders, make sure they understand the problem you are trying to solve and the business impact that would occur if they do not solve the problem. Quantify in specific dollars the benefits of solving the problem with mobile learning. Let's look at a couple of examples.

Instead of saying you'll reduce the costs of your technicians in the field by reducing the time to fix common problems, quantify how much money could be saved by giving them the ability to troubleshoot and fix routine problems faster.

Instead of saying you'll reduce the number of safety incidents in the field, quantify how much money could be saved by reducing the number of safety incidents. Use hypothetical models. For example, say, "If we reduce the number of safety incidents by 15 percent, we would save $X next year."

Instead of saying you'll increase sales, quantify how much could be earned if the sales team could close X more sales a quarter.

A key to gaining executive stakeholder support is to paint the picture that going forward with the mobile learning project makes financial sense.

What Is Your Plan to Evaluate the Program?

Once your executive stakeholders understand the business problem you will solve with your mobile learning solution and they realize exactly how this will result in decreased costs or increased revenue, you must address the following questions: How will you measure the effectiveness of the solution? How often will you measure it? It is a good thing you thought this through early in the process when you drafted your mobile learning vision.

Do you remember the four ways we discussed in Chapter 5 that you could use to define and measure the success of your training program? As a refresher, here they are:

1. Employee satisfaction with the program in meeting their needs and expectations
2. A change or increase in employees' knowledge, skills, and attitudes based on their participation in the mobile learning program
3. A change in employees' on-the-job behavior
4. The business impact on your company

Business Impact Takes Time

SMART MANAGING

Your stakeholders will want to talk about and see the positive impacts that your mobile learning solution has had on the health of the business. However, it takes time to reap the benefits of any training program and to be sure that the results were due to your mobile learning solution, not other influences. With that in mind, discuss with your stakeholders how you'll determine that the program is on its way to being a success. This can be accomplished by showing them how your employees' behaviors are changing, which ultimately will result in the anticipated business results. Communicate with your stakeholders when and how often you plan to take the measurements and when, how, and how often you'll communicate the results to them.

Part of your success criteria will fall within the categories of measuring your employees' satisfaction with the program or measuring how much their new level of knowledge, skills, and attitudes changed due to the training experience. You should also address these as a way to determine if you're meeting your end-user stakeholders' needs and expectations. You'll probably spend most of your time addressing how exactly you'll measure the impact on your organization.

Since a change in results usually requires your employees to change a certain aspect of their behavior, I recommend that you start by measuring behavioral change. Explain the specifics of your plan including:

- How you'll observe the change in behavior and the frequency
- Whether you'll take a 360-degree look at the perspectives of the employees, their peers, and their managers
- Asking line managers to perform on-the-job observations
- The types of tools (forms, checklists, job aids, etc.) that line managers will need to perform the evaluations
- Documenting employees' on-the-job performance through some sort of technology

- Ensuring that the behavioral change is long term versus short term
- How often you'll perform the evaluations
- The type of feedback you'll provide to the employees on the extent to which they've changed their behavior.

Next, focus on explaining how you'll measure the results or business impact from your mobile learning solution. This should include:

- The metrics you'll measure
- Which systems will be used to analyze the data
- If you'll need to work with IT to combine data from multiple sources
- If you'll create a dashboard to track the progress
- How you'll baseline the current state
- How often you'll measure and when in the process
- Your plan to communicate the results to the stakeholders
- What level of stakeholder support (if any) you may need to pull together the necessary resources
- How you'll communicate your short-term and long-term measurement plans

An Approach That Minimizes Risks and Costs

Another key point to gaining buy-in from your executive stakeholders is to show them how your approach will maximize the return on investment while minimizing risks. An approach I recommend is to think big, but implement in small steps. What do I mean by that? I'm sure you have some grand ideas for how you can implement mobile learning.

Warning: do not roll out a complete program at one time, especially if your company is new to mobile learning. Your stakeholders may be skeptical about how this new approach will work in the field, and there are many uncertainties regarding technology and whether the employees will embrace the program. A phased rollout will not only allow you to simplify the process, but also allow you to get quick wins. It's better to deploy a small solution and take an opportunity to learn from incremental results. See what works well, identify where you can improve, and make adjustments as you go.

Another point to address regarding risk management is your game plan for ensuring that all stakeholder requirements are being met,

IDENTIFY LOW-HANGING FRUIT

TRICKS OF THE TRADE

I recommend that you look at the business problem you are solving and identify the low-hanging fruit that you can quickly address through mobile learning. Think of low-hanging fruit as a few scenarios that you can implement as a mobile learning solution that will have an immediate, positive impact on the company. Roll out that project to your employees first. Let them use that part of the solution and evaluate what works well and where it needs adjustment to work better. Not only will you minimize risk by limiting the scope of your project, but your organization can also quickly see a return on investment. This will give your executive stakeholders the confidence to move forward with additional mobile learning projects.

including learner needs. You accomplish this by providing opportunities to gather learner feedback, adjusting the program by prototyping your solution, and running a pilot program before investing in full-fledged development. In the next sections, we cover these points.

AN APPROACH TO OVERCOMING RESISTANCE

TRICKS OF THE TRADE

If you experience some stakeholder resistance to buying into the validity of your mobile learning concept, suggest that they invest in a proof-of-concept project. A proof-of-concept lets you test the waters with a small, targeted project before your organization invests in the full project. This gives you an opportunity to address stakeholder concerns or objections. Identify the resistance points and devise a small project that shows how the solution will work. Focus the project specifically to address the stakeholders' barriers to acceptance. Keep in mind, a proof-of-concept project should prove an idea or concept; it does not involve a complete evaluation and testing process. A proof-of-concept allows you to show a single aspect of the project to your stakeholders.

Eliciting Feedback from Stakeholders

You need to keep your stakeholders in the loop as to the health of your project. They should be given the chance to experience the solution as it is being built and provide their feedback.

Gathering feedback early in the process and frequently throughout the project is an important component of a successful project and helps to gain your stakeholders' trust and buy-in. Without finding ways to incorporate stakeholders' feedback, it is almost impossible to create a solution that meets all their requirements. Stakeholders do not necessarily know exactly what they need, nor are they always able to verbalize their needs, especially with a new concept such as mobile learning. Have you ever had someone tell you, "Oh, I know I said that, but that's not really what I meant"? Happens all the time in such projects. We often believe we are communicating clearly when, in fact, our words are open to interpretation.

Another issue arises with mobile learning, and that is creating an experience in an uncontrollable environment. While our stakeholders may have a notion of what the experience *should* be, until they have the opportunity to see the solution in action, use it in context, and confirm it, we are at risk of missing the mark. As a technique to gain stakeholder buy-in, I recommend that you take the approach of "show, don't tell."

STAKEHOLDER FEEDBACK

When your stakeholders review the deliverables, obtain objective feedback from them that your team can act on. However, it's likely that you'll receive a number of subjective comments, typically starting, "I don't like . . ." While this is good information to know, you need to turn those subjective comments into actionable items.

You do this by asking follow-up questions about why they do not like a particular element or feature and push them to articulate a solution. If your team acts on subjective feedback, you'll find yourself in an unending loop of making changes until everyone can agree on what they like.

Key Deliverables That Minimize Risks and Costs

There are several deliverables you need to know about that can reduce costs and minimize risk of failure.

Wireframes

One area of risk when developing mobile learning lies in designing a user interface that's easy to use and runs on a wide variety of devices. To minimize the risks associated with interface design and usability, use a wireframe. Think of a *wireframe* as the skeleton for the design and function of the mobile learning interface. Wireframes let you show your

KEY TERMS

Wireframe A visual tool that shows the structure, functionality, and information architecture for your mobile learning experience. It doesn't contain the final graphical images, instead showing the "bones" of the program and how it will work. Also known as *paper prototypes, blueprints*, or *page schematic.*

Prototype A physical model of the learning solution that lets you see and/or experience firsthand how the learning program will be designed or function. Depending on the prototype goals, the model can be anything from a paper representation of the graphical user interface, to a technology-based, click-through example showing how specific techniques and interactions will work, to a fully functional piece of software that allows you to grasp a portion of the total experience.

stakeholders an early draft and gain their feedback and guidance on:

- The user interface elements, their location on the screen, and what they do
- The organizational structure of the application and content
- The functional elements and how they will work
- How the learner will interact with and use the application

SMART MANAGING

Wireframes Are Not Pretty

One challenge you may face when walking your stakeholders through your project's wireframes is that they are not eye-catching. Wireframes are basically black and white block diagrams of each screen that display all the elements and indicate how each will function. Start your review meeting by explaining that wireframes are intentionally devoid of all graphic elements (such as colors and images) to allow them to easily focus on the larger picture. Reassure stakeholders that they will have an opportunity to see these elements when the wireframes are brought to life in the prototype.

Once you and your team have defined the functional requirements and have developed some use cases, the designers use the wireframes to generate the first major design deliverable. Wireframes may be sketches the designers have mocked up on paper, created using PowerPoint or Keynote, or developed with a wireframe software program. Wireframes are not functional; they allow you to follow screen by screen how the program will function.

Ask for Critical Feedback

I recommend that you walk your stakeholders through the wireframe deliverable. A walk-through could illuminate any confusion as to the solution's intent and can elicit critical feedback from your review team. Structure this review meeting as an interactive experience. Ask questions, seek feedback, and encourage your stakeholders to discuss all the elements of the wireframe. Ask probing questions to obtain critical and meaningful feedback. Questions to ask include:

- Is there anything missing from the user interface?
- Are there additional functions or user capabilities not shown in the wireframes?
- Do the stakeholders have concerns about any of the wireframe elements?
- Does this initial look at the program meet all their requirements?
- Do they see possible usability issues?
- Do all the stakeholders agree on what will be built and how it will function?

Your goal at this review meeting is to ensure the product meets all the business specifications and the user experience and functionality requirements before you develop your solution. This is an easy point in the project to make adjustments, and you should expect that your stakeholders will bring up some important points that had not been previously considered.

This deliverable is a key review point for you and your project stakeholders. Your stakeholders can see the project in a structured, organized format without spending money on programming.

Prototypes

Mobile learning is new to organizations, hence teams encounter a number of unanticipated challenges that their designs must account for in order to create an effective mobile learning solution. Simulating and going through your employees' user experience let you evaluate specifically what works and where changes are needed. By creating a prototype, you gain the opportunity to test your team's design ideas, try out new functionality, and test technology elements before you invest in fully developing your solution. Using a prototype takes the guesswork out of the program.

Creating a prototype is beneficial because it:

- Increases communication between the stakeholders and the project team during the design phase

> **WHAT TO PROTOTYPE**
>
> SMART MANAGING
>
> There is no right or wrong answer when it comes to what to prototype or how much of your solution you need to prototype. Your company's experience with mobile learning in the workplace will affect what you prototype. For example, if you have already deployed a solution with similar functionality, you probably will not have to prototype that solution. A general rule on what to prototype falls into three categories: look and feel, content, and functionality. You may also find that you invest in multiple prototypes. For example, if your team has two different ideas on how to deliver the content, you may want to create two prototypes to determine which approach works better for your end users.

- Allows you to quickly adapt to changes in program requirements
- Allows you to quickly gain critical feedback
- Encourages collaboration and freedom to test new ideas or approaches
- Reduces the costs of the program by testing and updating the solution before developing it
- Minimizes project risk by finding what works and what does not

Before your team creates a prototype, you want to make sure there is a defined goal for the prototype. What exactly will you evaluate? Next, you should determine how much of the program should be modeled so your team can evaluate it. Remember, prototypes are only a part of your solution.

The next point to consider is the level of realism required to experience and evaluate the prototype and what tools the design team will use to create it. In some cases, a paper-based version of the solution is enough; in other cases, you'll need a functional model. Functional models do not mean that your team will be coding the solution. The team could use a software program such as Word Press that quickly outputs HTML without requiring the team to write a single line of code, or the team could use PowerPoint or Keynote. You also need to think through how many iterations of the prototype you'll run. You may start with a sketch, gain feedback on it, then make your updates in a prototype that more closely resembles the user experience.

TRICKS OF THE TRADE

Hold a Stakeholder Team Review Meeting

Design teams will often distribute a prototype for stakeholders to review. The stakeholders are asked to review the prototype and comment to the development team on the design. Stakeholders have busy schedules, and it can be difficult to obtain their feedback. However, not asking your stakeholders to provide feedback is a recipe for disaster.

Two issues that can arise from this process include conflicting guidance from multiple stakeholders and subjective feedback. A third complication involves making changes that may conflict with another stakeholder's requirements, who is unaware of the changes and has not agreed to them. To avoid these disasters, I recommend you send your prototype to all the stakeholders, giving them time to review it and document their thoughts. Then call a stakeholder meeting to go over the prototype and their feedback. Ask follow-up questions regarding their comments, make firm decisions on the action items, and gain group consensus.

Manager's Checklist for Chapter 9

- ☑ Success in gaining executive stakeholder buy-in lies in your ability to communicate a compelling business problem and to show how your mobile learning solution will increase organizational revenue or decrease costs.
- ☑ Your executive stakeholders will want to understand your approach to designing and developing your mobile learning solution and how your approach will minimize risks and reduce the costs of developing the solution.
- ☑ Quantify the expected results as a business value to the organization. Use hypothetical yet realistic models to paint a compelling picture.
- ☑ Have big dreams about mobile learning but execute in small chunks so you can minimize risk and development costs of the mobile learning solution, while providing the organization with quick wins and business value.
- ☑ Prototypes reduce risks and costs as they allow you and your stakeholders to experience the mobile learning solution without investing in programming costs. Modifications can be easily made at a low cost.

Working with the Development Team

Now that you have an understanding of the roles and responsibilities of your development team, let's look at the content creation process and the key milestones that require your involvement. There are two distinct mindsets when it comes to developing content. One is to use a traditional waterfall methodology called ADDIE that's commonly used by Learning and Development teams. The second approach is to use an agile software development methodology typically used by software development teams. Toward the end of this chapter, I also provide you with some tips on supplying feedback to your team.

Using ADDIE

When you speak to your Learning and Development team about how they design and develop content, chances are they'll talk about the ADDIE model. This model has been a staple for training professionals for years, and it's the way many ILT and e-learning courses are designed and developed. ADDIE is a systematic approach to course building and consists of five phases: analysis, design, development, implementation, and evaluation. During each phase of the process, the team performs a formative evaluation and adjusts the content as necessary. At the end of each phase, an approval process is required before moving on to the next phase. By looking at Figure 10-1, you can see how the processes work

Waterfall methodology A linear approach to developing training solutions where one phase must be completed and approved before the development team can begin working on the next phase of the project.

ADDIE An acronym for the process that many training professionals use to design and develop training programs such as e-learning and classroom training. The process consists of five phases: analysis, design, development, implementation, evaluation.

Agile A group of software development methodologies that use an iterative and incremental approach to developing software. The process involves cross-functional teams in a highly collaborative environment.

Formative evaluation An evaluation method that reviews and validates whether the goals have been met at each phase of a project. Include a few target learners on your team to gain the best input on your formative evaluations.

together. Some argue that ADDIE isn't a waterfall approach because the team performs evaluations throughout the process and at each phase, multiple iterations can take place. In fact, you'll see many diagrams that show the phases as a cycle, with evaluation lines connecting each phase. However, this doesn't mean that the process allows for new or changing requirements. At the end of the day all your content is being deployed at one moment in time.

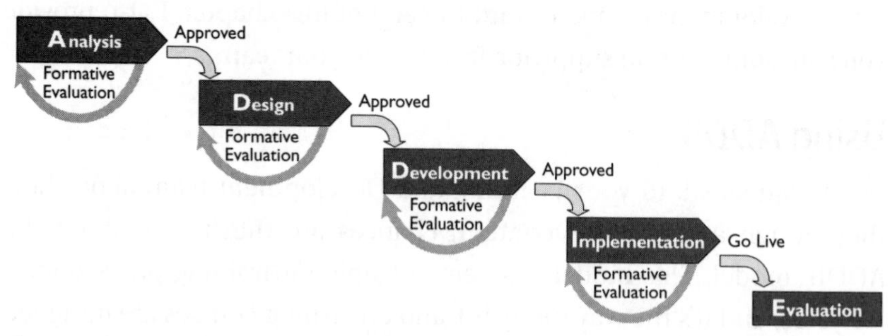

Figure 10-1. The ADDIE process

The Five Phases of ADDIE

Let's look at each of ADDIE's five phases to give you a better understanding of the process. We discuss the intent of each phase, key tasks that are typically performed, and the team members involved.

ADDIE MODIFICATIONS

CAUTION

If your company has an established Learning and Development team, chances are good that they've modified the tasks that take place within each phase of ADDIE to develop their own version that works well for your company. When working with a development team that uses ADDIE, find out exactly if and how they've modified the process. Make sure that you have a good understanding of the key tasks that the team performs at each step as well as the documentation the team provides.

Phase 1: Analysis

Analysis is the most critical phase. The intent of this phase is to minimize risk by thoroughly analyzing all aspects of the course before investing in designing and developing the solution. This phase is primarily performed by the instructional designer who:

- Identifies the problem and determines if it can be solved by providing training
- Analyzes and establishes the business goals for the training
- Determines the objectives the program must achieve
- Performs a detailed task analysis
- Analyzes your audience's current knowledge, skills, and attitudes compared to the desired state, and determines the gap between the current and the desired future states
- Reviews existing materials
- Determines how to measure the effectiveness of the training program
- Analyzes the technology environment and determines the available delivery options

The output at this point is called a *needs analysis document* and must be reviewed and approved by the project manager, subject matter experts, and stakeholders. This document commonly goes through multiple iterations prior to approval.

Once the needs analysis document has been approved, the next ADDIE phase is to document the training solution for the problem. This document includes a detailed description of the recommended training solution, including delivery method(s), total seat time, project timelines, project risks, and project costs. This document must also go through a

Seat time The anticipated time during which the average learner will interact with the course to complete it. For example, when we attend an 8-hour, instructor-led course, the average seat time is 6.5 hours, allowing for a 1-hour lunch break and two 15-minute breaks.

review and approval phase.

Phase 2: Design

In the design phase, your team documents the training program's design elements, including all the specific content to be covered, performance and learning objectives, look and feel, and how media elements will be incorporated. Your instructional designer and graphic designer are the primary resources for this phase. However, a developer may be called in as a technical expert. The instructional and graphic designers:

- Identify the specific learning objectives for the program
- Determine your assessment strategy and write the assessment questions
- Organize and outline all the supporting content
- Document how you'll use visuals and instructional strategies in the training program
- Document how the technical elements will work
- Design the user interface and explain the function of all interface elements

The final deliverable at this stage is a detailed design document that again requires review, updating, and final approval by the stakeholders and your review team. Once the design document is approved, the next step is

Don't Skip the Prototype Step

Many years ago when e-learning was new, it was common practice to invest in a prototype; however, today many e-learning teams bypass this step. Since mobile learning is a relatively new field of practice and with all the complexities of designing for multiple devices and operating systems, it's imperative that you create at least one prototype of your proposed solution before moving into the development stage. A prototype reduces costs by addressing the unknowns early in the process and provides your team with insight regarding what works and doesn't work with your proposed solution.

to design a prototype that brings the solution to life before moving on to the development stage. The prototype must be reviewed and approved by the project manager and stakeholders. You may also want to include key subject matter experts in this process.

Phase 3: Development

In the development stage, the team focuses on bringing all elements of your learning solution to life and tests the solution. This includes writing the lesson content, creating the media, and programming the content and integrating the solution with other systems such as an LMS. At this point in the project all your team members are actively participating by:

> **Storyboard** A visual script that instructional designers create that shows page by page the screen text, the narrator and character audio, a description of the images and/or illustrations, and an explanation of the interactions and animations.
>
> KEY TERM

- Creating storyboards for all lessons
- Creating images and/or holding photo shoots
- Recording audio files
- Performing all lesson programming and animation effects
- Adding SCORM calls to lesson content
- Reviewing all lessons and testing the program

At the end of this step, lessons are reviewed and adjustments are made as needed prior to final approval.

Phase 4: Implementation

In the implementation phase, you perform the final preparation steps

> TRICKS OF THE TRADE
>
> **Approve Storyboards Before Programming Begins**
>
> As stated earlier, the ADDIE process is often modified. Some companies create the storyboards in the design phase. For those of us who wait until the development phase, I recommend that you add a second approval step: have your storyboards reviewed and approved prior to moving forward with the development. This additional approval step saves you time and money by not having to reprogram your content. And depending on the type of solution you are creating, you may want to skip the storyboards and use wireframes as a better solution, as happens with many mobile learning apps.

and go live with your training solution. That includes:

- The final content publishing
- Ensuring LMS is set up and ready for users
- Educating your employees on the program and student registration
- Going live for the users

Phase 5: Evaluation

Throughout each phase of the process, the team performs a formative evaluation; however, at this stage you're focused on performing a summative evaluation. Remember back in Chapter 5 when we spoke about how you'll evaluate the effectiveness of your solution? Well, this is the stage where you perform the formal measurements and determine if you were successful in solving your business problem. This phase also allows you to see where you must further adjust the program. Key areas that you may assess include:

KEY TERM

Summative evaluation Evaluating the effectiveness of your training program once it has been used by the learners.

SMART

MANAGING

YOU MAY NEED ADDITIONAL ASSISTANCE IN EVALUATION

Many training organizations lack the money or direct access to the resources required to evaluate the effectiveness of their mobile learning at all four levels. In fact, when I talk to training teams about the evaluation step, they often tell me that once a program goes live, they don't get involved until it needs to be updated or modified. That said, most companies do perform a Level 1 survey (the "smile sheet") to measure the users' reactions. Many organizations also perform a partial Level 2 evaluation based only on test results, which really only measures recall and not necessarily an increase in knowledge or skills and attitude change. Both of these are often features of the Learning Management System, so that measurement is easy to implement.

Unfortunately, many training organizations don't truly measure if there was an overall change in behavior or if the solution had an impact on solving or addressing the business problem at hand. Not all solutions require you to measure all four areas; however, if you need to measure behavior change and business impact, you may need to look beyond your training department. You may need to take the lead in working with IT on collecting pertinent data or setting up a dashboard. Another option is to partner with a vendor that can help you with the process.

- Learner satisfaction with the training program
- Any increase in learners' knowledge
- Changes in employees' behavior on the job
- Overall business impact

The Pitfalls of Using ADDIE

I've personally used ADDIE for almost two decades, and it can be a good process. However, don't go into the process with your eyes closed, as it has some significant pitfalls.

Since requirements must be defined and locked down early in the process, many companies spend more time than originally estimated on the analysis phase and experience "analysis paralysis."

Once a phase is approved, it's difficult to go back and make adjustments. This is why the process is called a *waterfall*. Have you ever seen a waterfall go backward? Of course not; once the water moves down a level, there's no going back.

This process limits creativity. Once the storyboards have been approved, the development team tends to bring them to life exactly as approved. Unfortunately, many organizations have their instructional designers and developers working apart from each other, and they don't collaborate on the storyboards.

Testing occurs when all the final lessons are created. This can cause significant rework. For example, you find your interactions are not meaningful to the learners, but you've already developed them. If you have a coding problem, then you need to rework all the lessons.

You don't deploy the curriculum to learners until it's entirely developed. This process can take a significant amount of time. I've personally managed very large projects that took up to 18 months to go from the analysis phase through the implementation phase. Not only do your employees have to wait a long time for their training, but also quite a bit will have changed in the organization during this time.

Key Project Milestones That Need Your Involvement

When using the ADDIE model you need to be an active participant in all the formal review processes. I recommend that you hold a separate review of each step and provide feedback to the development team prior to releasing them to the stakeholders for their input.

The key milestones include:

- Needs assessment and solution description documents
- Design document
- Prototype reviews
- Development review
- Testing
- Evaluation of the solution

SMART

MANAGING

TEST YOUR MOBILE SOLUTION

Okay. So this phase may be met with some resistance, but I believe that managers have a unique perspective when it comes to testing solutions. I don't expect you to find all the bugs and issues, but you'll probably identify some issues the team overlooked. On many occasions, my team has asked, "How did you find that problem?" My response has been, "I don't know. I was just using the device as I normally do in my daily life while I was testing."

Doing things like changing the orientation of the device, opening up a second webpage or a different app, turning on the airplane mode just as I would when I travel, or setting the device down while I'm testing to work on other items and then having to unlock the device. These are just things I do whenever I'm testing and, amazingly enough, it always seems like I run the mobile learning through a few scenarios that others didn't consider.

Getting Agile with Your Development

While many training and development professionals use ADDIE to develop mobile learning solutions, you may want to look toward an approach that software developers and IT groups use called agile software development. While there are several agile methodologies to choose from (i.e., Scrum, Extreme Programming, Adaptive Software Development), they're all guided by core values that are explained in the Agile Manifesto.

The Agile Manifesto was written by a group of individuals who have used a wide variety of software development methodologies. These developers shared the common goal of wanting to establish some key core values and principles to aid the software development industry in creating higher quality solutions in a shorter time. This group of practitioners founded a nonprofit group, the Agile Alliance. I quote The Agile Manifesto below, retrieved from the Agile Alliance website (www.agilealliance.org/the-alliance/the-agile-manifesto).

Agile Manifesto

We are uncovering better ways of developing software by doing it and helping others do it. Through this work we have come to value:

- Individuals and interactions over processes and tools
- Working software over comprehensive documentation
- Customer collaboration over contract negotiation
- Responding to change over following a plan

That is, while there is value in the items on the right, we value the items on the left more [i.e., the words before the word *over*].

Twelve Principles of Agile Development

TOOLS

Following are the 12 principles that agile development methodologies leverage as stated on the Agile Alliance website (www.agilealliance.org/the-alliance/the-agile-manifesto/the-twelve-principles-of-agile-software). These principles expand on the four key values of the Agile Manifesto.

1. Our highest priority is to satisfy the customer through early and continuous delivery of valuable software.
2. Welcome changing requirements, even late in development. Agile processes harness change for the customer's competitive advantage.
3. Deliver working software frequently, from a couple of weeks to a couple of months, with a preference to the shorter timescale.
4. Business people and developers must work together daily throughout the project.
5. Build projects around motivated individuals. Give them the environment and support they need, and trust them to get the job done.
6. The most efficient and effective method of conveying information to and within a development team is face-to-face conversation.
7. Working software is the primary measure of progress.
8. Agile processes promote sustainable development. The sponsors, developers, and users should be able to maintain a constant pace indefinitely.
9. Continuous attention to technical excellence and good design enhances agility.
10. Simplicity—the art of maximizing the amount of work not done—is essential. [continued on next page]

11. The best architectures, requirements, and designs emerge from self-organizing teams.
12. At regular intervals, the team reflects on how to become more effective, then tunes and adjusts its behavior accordingly.

I list below some of the benefits of an agile approach over a waterfall methodology, such as ADDIE, specific to mobile learning.

- Both the learners and the company benefit from learning and performance support solutions being deployed more quickly.
- Companies can quickly begin to generate a return on their investment.
- An agile approach gives development teams an opportunity to learn from what works and what doesn't, and apply those updates to the next release.
- Developers don't always know all the project requirements until they get further into the process, and sometimes requirements change.
- An agile approach fosters a collaborative development environment where the team takes on project ownership and constantly evaluates how the product and the process can be improved.
- Frequent communication between the development team and the project stakeholders ensure expectations are being met.
- Agile methodology gives developers a tremendous amount of flexibility to deal with the unknown.

Now that you have an understanding of the values and principles associated with agile software development as well as the benefits specific to mobile learning, let's look at Scrum (see Figure 10-2).

Scrum

Let me walk you through the Scrum process. You'll see in Figure 10-2 that the project begins with a vision. Does this sound familiar to you? We discussed the importance of creating your mobile learning vision in Chapter 5. Once there is a vision for the project, the team begins to plan the project. This is a collaborative approach in which the product owner, Scrum master, and team brainstorm to identify all the program requirements and features. The result is a list of high-level user stories that describe in a couple of sentences who the users are, what they'll be doing, and why they'll be performing that function (or for what end result). The team creates an initial list of user stories. (A *user story* is a high-level explanation of all the mobile learning project functionalities broken down into unique features and functions of the software. The stories consist of the name of the feature and a brief description.) The product owner prioritizes them, listing the most important user stories first. This is called the *product backlog*.

Scrum One of the most widely used agile development methodologies. Scrum uses its own terminology to describe the process whereby software is developed and deployed through a series of small iterations called *sprints*.

Product owner The person who has the overall vision of the project and has the ultimate responsibility to ensure the team provides value to the organization. While from this definition you may think you are the product owner, this is a full-time tactical job. Think of yourself as the strategic product owner or the team's customer.

Scrum master The person who acts as a facilitator or coach for the product owner and the team. For example, the Scrum master works with the product owner on how to best apply Scrum to maximize the owner's return on investment. The Scrum master also works with the team to apply Scrum principles as well as work through problems that may cause the team to not meet their goals.

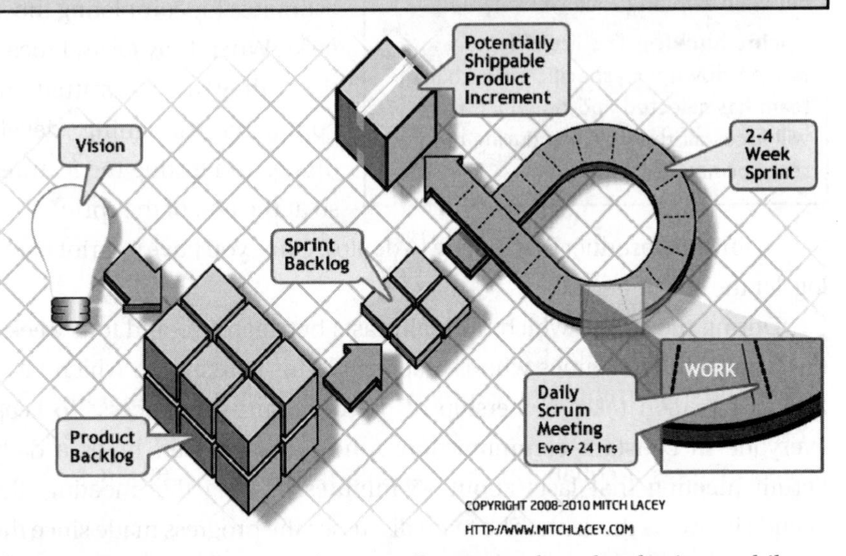

Figure 10-2. Scrum framework—an agile approach to developing mobile learning

At this point, you've completed the first planning step. Instead of taking months, which the analysis step could take in the ADDIE model, you have an initial list of requirements within days. Don't worry that the team may not have accounted for all the possible features and requirements. One of the benefits of being agile is that you can add to the product backlog as you identify new features. Now that your high-level planning is complete and you've created your product backlog, you can move into your first sprint.

The first step in starting any sprint is to hold a sprint planning meeting, for which a full day is normally allocated. During this meeting the product owner describes the highest priority features from the backlog to the team and answers questions the team poses. The team then meets to define the sprint backlog, which consists of defining a goal for the sprint, deciding which features they'll focus on to meet their goal, and creating estimates for completing those tasks. When I say *focus,* I mean that they're committed to completely designing, developing, and testing the features so at the end of the sprint, they have a finished product that's ready to deploy. Once you have a sprint backlog, let the sprint begin!

Product backlog A prioritized wish list of all the features and functionality that will be included in your mobile learning project.

Sprint A period of time allocated to developing, testing, and revising a set of functionality so there is a working product when the sprint is complete. Sprints can range anywhere from one to five weeks; however, they typically range between two and four weeks.

Sprint backlog The list of features broken down into specific tasks that the team has selected and on which they will focus all their efforts during the current sprint.

During the sprint, which normally lasts between two and four weeks, the team self-determines who is responsible for working on which tasks and that person takes ownership of the development process. To keep everyone in constant communication, the Scrum master holds a daily Scrum meeting that lasts about 15 minutes. During this meeting, the agenda is always the same. The team discusses the progress made since the last meeting, what they're working on today, and the challenges they're encountering. The Scrum master is responsible for ensuring the meeting stays short and on task, and "sidebar" conversations that need a more in-depth conversation are held later.

User Stories

TRICKS OF THE TRADE

To move away from long formal requirements documents, user stories document the requirements in many agile processes. Many groups use note cards to document the user stories. User stories flesh out the detailed requirements by having conversations with the team and the customer. When prioritizing a list for the product backlog, decide which user stories would generate the greatest return on investment, since Scrum can have multiple builds or releases of the software.

Upon completing the sprint, two meetings are held. The first meeting is the sprint review where the team goes over the work they've completed and provides a demo to the product owner and stakeholders. During this meeting, new requirements are identified for the project. These requirements go on the product backlog, and the product owner reprioritizes the list. The second meeting is the sprint retrospective, where the team openly discusses how to

Tips for Success with Scrum

TRICKS OF THE TRADE

Scrum is a new mindset for many individuals. With that in mind, here are a few tips to help make your project a success.

- First, the team chooses what they'll focus on in each sprint, not you. They best understand what can be accomplished and ready to deploy in the designated time frame. Let them control this element of the project.
- Typically, the team determines who is responsible for each component of the solution. A project manager does not have to perform this role.
- The product owner's role is to ensure your team understands the project's vision and to prioritize the functions to provide business value. By providing guidance to the team on the most pressing areas on which to focus their time, you are managing the project to reach its goal.
- A key to prioritizing the product backlog is to identify those requirements from which the company would receive the greatest return on investment.
- Once the team has started a sprint, don't under any circumstances add requirements. Let the team focus on what they've committed to finishing.
- Use the sprint goals agreed on in the sprint planning meeting as the basis for progress reports to the stakeholders. They don't need to know about every piece of functionality.

improve the overall mobile learning product and the development process. The process continues by beginning the next sprint planning meeting.

After a sprint or two, you should have a product that's ready to release. It's probably not a complete package, but you'll be able to benefit from releasing some level of functionality to your workforce. You'll also be able to get feedback and perform evaluations of the solution instead of waiting until the complete product is ready.

Key Project Milestones That Need Your Involvement

If your team is using Scrum and the project is not too large, you may take on the role of the product owner. With that in mind, here are the key points of the project that need your attention.

- Creating the project vision
- Managing and prioritizing the product backlog
- Participating in the sprint planning meetings
- Attending and actively listening during the daily 15-minute Scrum meetings
- Providing feedback at the sprint review meetings

If you have assigned a different resource person to be the product owner to take on the role and daily responsibilities for the project, you:

- Communicate the project vision to the product owner
- Provide input on user stories, managing and prioritizing the product backlog
- Hold weekly meetings with the Scrum master and the product owner to discuss the project
- Participate in and provide feedback during the sprint review meetings

Creating a Hybrid Approach

Both ADDIE and Scrum have their strengths and weaknesses and, honestly, either approach is fine. In fact, some companies adopt the best of both approaches and create a custom process. Speak to your team on how you can create a process based on iterations where you can learn what works and what doesn't, and quickly make adjustments. Let me give you a few ideas to get the conversation rolling with your team.

Within your mobile learning vision, you have already spent time documenting many of the areas your team will analyze in the analysis phase. This includes:

- Identifying the problem and determining if it can be solved by providing training
- Analyzing and establishing the business goals for the training
- Determining the objectives that must be achieved for the program
- Determining how to measure program success

Since you've already written the initial documentation, have your team confirm these assumptions rather than perform a detailed analysis. With mobile learning, you need to be flexible and allow for adjustments in your requirements and even your approach. Don't feel like the team must know everything up front.

Next, have your instructional designers determine the tasks and content that need to be covered to meet your goals and identify what content already exists.

Have your team brainstorm the design elements and requirements for your solution. Create a prioritized task/feature list and select some low-hanging fruit on which to focus your initial efforts. Think product backlog. Then start creating wireframes and prototyping this component to your solution.

When it comes to developing and implementing, use the concepts of sprints. This allows you to get a solution to your learners quickly, gain their feedback, make adjustments as necessary, and start gaining a return on your investment.

Do What Works for Your Team

CAUTION

When it comes to your development process, it's most important to use or create a process that encourages collaboration, creativity, and flexibility. Find a process that provides lots of opportunities to test and gain feedback from your learners and make adjustments as needed. While many mobile learning vendors use agile processes, don't feel that you need to apply the pressure of learning how to apply a new process such as Scrum to your team's responsibilities. If you are using your in-house team, ask them to look at how they develop content today and identify opportunities to tweak it to be more iterative.

Tips for Guiding Your Team and Providing Feedback

When managing your project team, here are a few tips for providing guidance as well feedback. This section highlights key meetings and communication tips.

Daily Updates

Regardless of the development methodology you use, I recommend that you hold daily team meetings. They don't need to be lengthy and should focus only on the high-level results the team has accomplished, what they'll work on next, what the risks are to completing their tasks, and if they need assistance to complete the tasks. This approach creates a team that grows more collaborative and avoids the "throw a task over the fence to another group" syndrome. It may seem like a burden to meet with the team each day, but it will not only ensure your project meets its goals but will show the team that you are actively involved and you support them.

Weekly Status Meetings

I recommend that the project manager send out a weekly status report on Friday afternoons. This report covers the progress made in that week and forecasts what the team will work on during the following week.

MAKING THE MOST OF YOUR DAILY TEAM MEETINGS

Here are a few tips to maximize the effectiveness of your daily team meetings.

- Schedule the meeting for the same time each day and for a set amount of time.
- Keep them short. This isn't a status meeting; it's a chance for the team to communicate with each other on the project's health and any challenges they are facing as a team.
- Encourage open communication.
- Require all team members stay through the end of the meeting. After all, everyone should be on the same page, and someone may bring up an item where the person who left could have been of assistance.
- Don't feel as if all problems need to be solved in this meeting. Use the meeting as an opportunity to identify risks and the team members who can help overcome them. Ask the team members to work out the detailed solutions away from this meeting.

The report also highlights any risks in meeting those goals. In addition to the written status report, I recommend a weekly meeting on Wednesdays for you to check in with the team. This halfway point meeting helps you and your team meet your current week's goals. It also gives you a chance to identify any challenges and brainstorm ways to overcome them.

Providing Feedback

Here are a few tips on providing feedback to your team that will foster a positive environment and ensure that your project goals are being met.

- Give clear directions on requirements.
- Be specific, and don't speak in generalities.
- Provide objective feedback that includes a solution.
- Avoid using the phrase "I don't like." This lends itself to subjective feedback and offers no solution.
- Make sure the team understands what you think you've communicated to them. With verbal communication, I often recommend having them explain in their own words what they've heard you say. Or you could ask them to create an action list.
- Frame your feedback from the position of the learners' needs or achieving the program's goals.
- Talk with the team about the problems you are trying to solve, and together, you may come up with a better solution than you could on your own.
- Compliment what is working well and when team members have met or exceeded your expectations.
- Celebrate your team's successes.

Consolidate Feedback

SMART MANAGING

When feedback is coming in from multiple sources, make sure that your project manager has reviewed and consolidated it prior to communicating it to the team. There is nothing worse from a team perspective than having to weed through a list of feedback points from multiple sources, organize it, and then determine what needs to be done. Make sure all feedback is consolidated by screen or page, and that you provide clear direction. If you have contradictory feedback, resolve it prior to handing it over to the team.

Manager's Checklist for Chapter 10

- ☑ Understanding the process of designing and developing your mobile learning solution is a key ingredient in working effectively with your team.
- ☑ Your Learning and Development team is probably using the ADDIE process, which consists of five phases: analysis, design, development, implementation, and evaluation.
- ☑ Agile methodologies that software developers use are more flexible in meeting potentially changing requirements, allow you to deploy content more quickly to your employees in a phased approach, and let you and your team evaluate at each deployment what's working and what needs adjustment.
- ☑ Scrum is an agile methodology that many groups use to design and develop mobile learning solutions. It's based on an iterative approach, consisting of a series of short sprints lasting two to four weeks. At the end of the sprint, you have features and functions of your mobile learning solution that have been developed, tested, and are ready to deploy to your employees.
- ☑ To increase communication among team members and identify potential challenges, meet daily with your development team for about 15 minutes.
- ☑ Hold weekly formal meetings with the team to discuss the health of the project and how you can provide assistance to the team in meeting the goals.
- ☑ When supplying feedback to your team, be objective, not subjective; give them consolidated feedback from the stakeholders; provide clear directions on solutions; frame your feedback in terms of meeting the overall goals or learner experience; and praise what the team has done well.

Ensuring a Successful Project

If you've followed the advice offered in this book, you're well on your way to managing an effective mobile learning solution that will generate positive results for your organization. While you cannot ensure a 100 percent successful project, I'd like to leave you with some tips that should improve your odds. This includes running your mobile learning solution through a pilot project and discussing ways that you can ensure your team is communicating clearly. Last, I provide you some tips on how to select your project manager.

Piloting Your Mobile Learning Solution

A key ingredient to ensure that your mobile learning solution will be a success is to test the waters with a pilot program. You may be thinking, "Isn't that why we develop prototypes?" To a certain extent, yes, that's true. But remember that a prototype is typically not fully functional and is part of the design and development process. A *pilot program*, however, is a scaled-down version of your mobile learning solution that is fully functional. By running a pilot you are able to experience exactly how your employees will interact with your mobile learning solution on the job. This is an opportunity to try out your solution on a small scale to determine what's successful and what you need to modify before rolling out a larger initiative. In this section, I give you some key points to con-

Pilot An initial rollout of your mobile learning initiative that's limited in scope (in content and number of users) to test the project's feasibility, identify technical issues, and evaluate the impact of the project.

sider when preparing for your pilot, and I also share some of the feedback and information you might receive when running a pilot program.

TRICKS OF THE TRADE

STAKEHOLDERS SHOULD USE THE PILOT PROGRAM

While not part of your target audience, you want to make sure that your project stakeholders have the opportunity to actively participate in the program. There's no better way to validate that the program has achieved its requirements and expectations than by having stakeholders experience it firsthand.

This is a great way to generate buy-in for funding the rest of the project and to generate high-level excitement about the potentials of mobile learning. Keep in mind that if you're providing the devices your employees will use, rather than allowing them to use their own, make sure that you provide your stakeholders with a device as well.

Preparing for Your Pilot

The first consideration for your mobile learning pilot is to define what a successful program looks like. You should also consider what metrics you will measure during the pilot phase. Next, determine the length of time that you'll run the program. On average, pilot programs run between one and three months. There is no magic number; however, you want the program to run long enough to give a good sample representation of typical usage and ensure that the data you collect is meaningful.

When preparing for your pilot, select employees to participate in the program. You need a sampling of your primary audience. You also want to ensure that you select people of varying skills with mobile devices. You don't want your pilot program to only go out to your mobile power users. In addition to selecting your program's active participants, select the individuals who will support your employees during the pilot. This may include help desk employees and representatives from your employees' peer group who have been involved in the prototype. You're probably

SELECTING THE RIGHT PERSONNEL FOR YOUR PILOT

MISTAKE PROOFING

When you design and develop your mobile learning pilot, your personnel choices are critical. Sometimes managers assign a less experienced person to this task to see if he or she has what it takes to work on a larger project. In my opinion, this is a bad idea. If you want to test less experienced people, then assign them to the final product team and assign mentors to work with them. The pilot is a critical review point, and you want to put your best foot forward. Another factor to consider in resourcing is whether your pilot team will be available to work on the final, larger-scale product. When I staff the pilot project, I tend to select the individuals who will take on the project's lead designer and developer roles to ensure that nothing is lost in a knowledge transfer and to ensure that a consistent product is developed.

changing the way they usually interact with training, and chances are at least a few of them will have questions. Last, log all the questions and issues that your pilot group encounters so you and your team can review them when the pilot is complete.

PREPARING YOUR EMPLOYEES

SMART MANAGING

Before going live with your pilot, educate your pilot participants on the project's intent and the role they will play in its success. Remember that you're changing how your employees interact and even think about corporate learning opportunities. Tell them how valuable their participation and input are. You must communicate the goals of the pilot, the length of time it will run, and how they'll provide their feedback to you. Also, if you're initiating a new device policy or acceptable use policy, take this opportunity to educate your employees on it and address their questions. Provide them with a frequently asked questions (FAQ) document and contact information for who will provide support if they encounter problems with either their devices or the learning experience.

Generating Buzz

You and your organization are embarking on an exciting journey by entering the world of mobile learning, and you should share that news with others in the organization. Before you launch the pilot program, ask a leadership champion to communicate why the organization is moving forward with this new approach and how the pilot participants are a critical component to the program's success. You could videotape an interview with this leadership champion that the employees can watch. In the video, include some brief demonstrations of the mobile learning solution. You could also write an article for the organization's newsletter and host a kickoff webinar.

In addition to leveraging your leadership champions, you should be talking to your peer champions about ways in which they can increase the buzz around the pilot program. You not only want to increase awareness of the program, but also generate excitement about your new mobile learning program. Ideally your peer champions were involved in the prototype phase and can share their experiences on how the program will make a difference to employees. If your organization has a social network already set up, use it as a way to connect with the employees.

What You Can Learn by Piloting Your Mobile Learning

Now that we've discussed some of the key points to consider when planning your pilot and generating buzz throughout the company, let's talk about the lessons you can learn from the pilot program. By running a pilot program, you gain valuable insights into:

- Your employees' reactions to the solution
- Whether your employees are using the program as anticipated
- How many people are taking advantage of the solution
- What is working well with the solution
- What elements need to be rethought or redesigned
- Employees' challenges with certain operating systems, platforms, or devices
- Whether the technology functions as anticipated

CAUTION

Don't Rush to Develop the Full Solution

A frequent mistake that managers and project managers make when running a pilot program is planning to develop the rest of the solution too quickly. I know you'll be excited to launch your full program and to start benefiting from your mobile solution; however, don't proceed until you've had time to analyze the early results. You and your team need to determine specifically what was effective and where you need to make adjustments, then create a plan to implement any changes to the solution as well as future rollouts. This can take some time, so make sure you plan for it.

- Whether the program is easy for employees to access and use
- What type of support calls you're receiving
- Whether you're successfully tracking all the data you need to measure a successful project

Ensuring Seamless Communications

Most of us would probably say that we are good communicators. Maybe we do, in fact, have that skill; however, most people have challenges in this area, and your project will require that everyone is communicating effectively. This starts at the beginning of your project when you clearly articulate the business problem you're solving and your goals for the mobile learning solution. Your stakeholders must be able to define their requirements and expectations for the project. Your team must be able to understand the guidance and feedback that are provided to them after deliverables have been reviewed. Communication is everywhere in your project, and finding ways to ensure everyone is on the same page is a critical component to a successful project.

Team Roster and Communication Preferences

When you begin your project, I recommend that you have your project manager develop a team roster that's shared with everyone on the team. The roster should include the individuals' names, role on the project, e-mail addresses, phone numbers, and preferred method of being contacted. This helps establish the importance of the complete team and encourages communications for the duration of the project.

The preferred method of communication is an important part of the roster. I've had many stakeholders tell me, for example, that they don't answer their phone and the quickest way to get a response is via e-mail. I've also had stakeholders explain that they get so many voice messages and e-mails that the best way to reach them is with a text message. To ensure timely responses, include that field on your roster, then make sure that your team understands the importance of using that method of communicating with that person.

Testing Log

As I discussed in Chapter 10, the testing process in a mobile learning solution is critical to creating a successful and effective product. However, testing is where a lot of miscommunication takes place. Assuming that your team uses an issue and/or bug tracking software program of some type, the communication issues only fall in two categories: identifying the location of the problem and providing all the information necessary to re-create and fix the problem. In what ways can you overcome these potential communication breakdowns?

MISTAKE PROOFING

DON'T USE A SPREADSHEET FOR TRACKING

A common practice for many Learning and Development teams is to use a spreadsheet to track the testing and quality assurance issues. Based on years of experience, I recommend that you don't do that; you'd be setting yourself and the team up for a disaster. These spreadsheet files must be uploaded to a server or sent out to others via e-mail. When people are using various versions of the spreadsheet or if a file gets loaded to the wrong place on the site or if someone is working on the spreadsheet on his or her local drive, then the file on the server isn't up to date. All these examples point to a communication hassle.

If you're going to be innovative about mobile learning, then invest in an issue tracking software program that allows real-time access to the issue log. In selecting the right software program to fit your needs, I suggest you meet with your IT department. They may have issue tracking software that will fit your needs.

First, when your team is designing your mobile solution, make sure they've created a naming convention so that each screen has a unique name. This allows your team to quickly identify where the issue is located in the context of the mobile learning solution and to quickly determine if the issue has been resolved. Second, to ensure seamless communication between the tester and the rest of the team, be sure you're capturing all the critical information. The following are some of the fields you want to make sure your team tracks:

- **Issue number:** a unique number that lets your team quickly identify and discuss a particular problem
- **Date:** the date the issue was identified
- **Version:** the version number or release of the software in which the issue was identified
- **Platform:** list the platform(s) as well as the operating system(s) on which the issue occurs

- **Description:** a detailed description of the problem and the steps to replicate it
- **Resolution:** the tester should provide a description of how to resolve the issue or the desired result
- **Attachment:** a field to allow your testers to add screenshots of the problem
- **Identified by:** the name of the tester who identified the problem and whom your team can ask additional questions on the issue
- **Assigned to:** the name of the person responsible for resolving the issue, and once resolved, the name of the tester who confirms that the issue has been resolved
- **Date due:** the date by which the assigned person must have the work completed
- **Priority:** a predefined ranking system that helps the team prioritize their workload to achieve deadlines
- **Date fixed:** the date the problem was resolved
- **Fixed in version:** in which version of the software the issue was fixed and deployed

APPROPRIATE USE OF THE ATTACHMENT FIELD

SMART CAUTION

The Attachment field allows your testers to take a screenshot of the problem and add it to the tracking system. This clarifies exactly where and what the problem is. The attachment cannot be used to replace documenting the steps needed to re-create the solution or any of the other fields. Think of it as additional information.

Communicating the Outcomes of Brainstorming Meeting

Throughout the book, I've suggested that you create an environment in which your team can work collaboratively to design and develop your solution and resolve problems. After all, when designing and developing mobile solutions, you need to think creatively about how to use the technology. Many benefits occur when team members take someone's idea and put another spin on it until the idea morphs into a solid solution. While this creative and collaborative environment is a good way to create solutions, it does present some communication challenges. For example, team members can easily lose track of the actionable items

DON'T ALLOW SILOS TO OCCUR

Sometimes, teams fall into the rut of silo thinking, which is when they stop thinking about the whole project and only think about it from the perspective of their individual area of expertise. For example, developers may consider their only responsibility on the project to be coding the solution. When allowed to work in silos, team members often leave status meetings when they've covered their tasks, or they tune out during meetings when the discussion does not cover a task that directly involves their role, or they do not actively participate in brainstorming sessions. This kind of attitude and behavior must be stopped immediately as it will have a negative impact, not only on the team, but also on the quality of your final product.

after a brainstorming meeting. To manage this risk, the project manager should provide a meeting debrief document that specifically describes the decisions the team will act on.

Weekly Meetings

Your project manager should distribute weekly status reports to all the key team members. In addition to that communication point, I suggest that you set aside a designated time each week to meet with your project manager and discuss the project. Use this meeting as an opportunity to delve into details than aren't included in the status report document. Find out if there are areas where you can be of assistance to the team in gaining access to resources that can help them overcome challenges, in working with other departments, or even in working with a hostile stakeholder. Also discuss with your project manager any team members who have gone beyond the call of duty so that you can personally recognize and praise their efforts.

Review Feedback

One area that may present your team with communications challenges is feedback from your stakeholders and/or review team. The first step I suggest is that, at each review point, you provide stakeholders and the review team with a review form and instructions for completing it. Set the expectations for specifically what they're reviewing. This will assist you in obtaining helpful feedback, but I need to point out that getting objective feedback with clear direction is a challenge. I suggest you hold your team's formal review meeting after your stakeholders and review team have had the opportunity to review the deliverables and document their feedback. Also, don't forget to include at least the lead development team resources to participate in this meeting as well as all your review team members. In this meeting, your team can hear firsthand what the issues are, identify conflicting feedback from various stakeholders, and come to an agreement on how to solve the problems.

Selecting the Right Project Manager

Choosing the right person to lead your team is an important decision. After all, he or she will be your right-hand person and responsible for delivering a project that meets your expectations, stakeholders' expectations, and user needs. When appointing a project manager, look for someone who has demonstrated:

- A high attention to detail
- Good communication and interpersonal skills
- Solid decision-making skills
- An ability to effectively utilize resources
- A knowledge of project management tools

I believe managing a mobile learning project requires some additional skill sets. When selecting a project manager for a mobile learning project, he or she should also:

- Be an active user of mobile technologies
- Focus on achieving your business goal and not a task list
- Be flexible when requirements change
- Have good presentation skills if he or she will be facilitating your review meetings

When a Vendor Provides a Project Manager

If you've decided to outsource the development of your mobile learning solution, the vendor will provide a project manager to oversee the process. I advocate that you designate an internal person to be your organization's project manager. He or she will be the point person for the vendor to interact with on a daily basis. Think of this person as a one-stop shop to resolve issues and ensure that your organization's goals and stakeholder expectations are being met throughout the process.

- Be an active listener to ensure he or she understands all of the stakeholder requirements

Manager's Checklist for Chapter 11

- ☑ Piloting gives you the benefits of rolling out your solution using the same process as a full project, giving you valuable insight into what works well, and if and what adjustments are needed.
- ☑ Key points to consider in the planning process include your goal for the pilot, what a successful pilot looks like, how long it should run, and selection and education of your participants.
- ☑ Generate buzz throughout the organization for the pilot program, share the project's goals, and tell users how the project will impact their jobs.
- ☑ Clear communication is vital to the success of your project, so take every step necessary to reduce barriers.
- ☑ Identify a project manager who has the skills necessary to effectively lead your project team and be your right-hand person.

Index

Index

G

H

I

J

K

L

About the Author

Brenda J. Enders is a consultant, trainer, public speaker, and author in the field of mobile learning. She is the President and Chief Learning Strategist of Enders Consulting, LLC, a St. Louis, Missouri-based company. Enders Consulting works in partnership with organizations in a variety of areas including devising strategies around mobile learning, game-based learning and gamification of learning, project management, and instructional design. Specific to mobile learning, she leads the visionary phase and manages solutions. These include mobile courses and assessments tracked within a learning management system, as well as supplementing or augmenting existing learning programs with mobile components to maximize learning retention and application, and lastly as a performance support tool to assist employees at the moment of need.

Prior to founding Enders Consulting, Brenda was the Chief Learning Strategist and Learning Services Practice Leader for a custom learning solutions provider for 16 years, where she led the design and deployment of innovative and award-winning custom learning solutions in partnership with Fortune 1000 clients and the U.S. government.

Brenda is a frequent speaker, workshop leader, and panelist at national and international conferences, as well as for corporations. Some of her presentations include: "The Next Generation of Mobile," "Mobile Training Reality: Employees Empowered by Learning," "Improving Performance through Gamification and Mobile Technologies," "Leveraging Cross Platform Development Without Sacrificing Learning Engagement in Mobile Solutions," and "Attracting and Retaining the Next Generation with Mobile Learning."

Brenda is an active member of The eLearning Guild, Serious Games Association, ASTD, ISPI, and CLO Business Intelligence Board. Brenda can be reached via her company's website at www.BrendaEnders.com, and you can follow her on twitter at @BrendaEnders.